Management and Industrial Engineering

Series Editor

J. Paulo Davim, Department of Mechanical Engineering, University of Aveiro, Aveiro, Portugal

This series fosters information exchange and discussion on management and industrial engineering and related aspects, namely global management, organizational development and change, strategic management, lean production, performance management, production management, quality engineering, maintenance management, productivity improvement, materials management, human resource management, workforce behavior, innovation and change, technological and organizational flexibility, self-directed work teams, knowledge management, organizational learning, learning organizations, entrepreneurship, sustainable management, etc. The series provides discussion and the exchange of information on principles, strategies, models, techniques, methodologies and applications of management and industrial engineering in the field of the different types of organizational activities. It aims to communicate the latest developments and thinking in what concerns the latest research activity relating to new organizational challenges and changes world-wide. Contributions to this book series are welcome on all subjects related with management and industrial engineering. To submit a proposal or request further information, please contact Professor J. Paulo Davim, Book Series Editor, pdavim@ua.pt

More information about this series at http://www.springer.com/series/11690

Carolina Machado · J. Paulo Davim
Editors

Entrepreneurship and Organizational Innovation

Springer

Editors
Carolina Machado
Department of Management, School of Economics and Management
University of Minho, Campus Gualtar
Braga, Portugal

J. Paulo Davim
Department of Mechanical Engineering
University of Aveiro, Campus Santiago
Aveiro, Portugal

ISSN 2365-0532 ISSN 2365-0540 (electronic)
Management and Industrial Engineering
ISBN 978-3-030-19288-4 ISBN 978-3-030-19289-1 (eBook)
https://doi.org/10.1007/978-3-030-19289-1

This Springer imprint is published by the registered company Springer Nature Switzerland AG
The registered company address is: Gewerbestrasse 11, 6330 Cham, Switzerland

Preface

In order to be competitive in nowadays markets, organizations need to develop innovative new business. Understood as the process of designing a new business, entrepreneurship has a close relationship with innovation. Indeed, concepts such as creativity, innovation, entrepreneurship, leadership, management ability, team building are critical issues in organizations that aim to be competitive. At the same time, due to the evolution that we have been facing in these scientific areas and to the fact that competition and the growing difficulties in entering into new markets are critical to business grow, modern organizations have been giving a special attention to innovation and intra-entrepreneurship as factors that make the difference in the development of their activities. Deeply present in competitive organizations, innovation has a great relation with entrepreneurship which, initially focused in the creation of new businesses, has been giving, in the last times, a particular emphasis to the organizations themselves, namely in what concerns the activities new profiles and tasks developed by workers. Understood this way, entrepreneurship also refers to the work developed inside organizations, where managers/entrepreneurs are seen as innovation agents who promote organization's continuous evolution deeply related with changes in the organizational culture as well as in tasks and workers profile, contributing to an innovative and entrepreneurial organizational environment.

Conscious of this reality, this book, entitled *Entrepreneurship and Organizational Innovation*, looks to provide support to academics and researchers, as well as those that operate in the management and engineering fields need to deal with policies and strategies related to organizational competitive issues. Presently, we live a time in which there is a vast amount of uncertainty and ambiguity about the direction in which organizations are moving. The need to reinvent new markets and new organization work models highlights the relevance that the entrepreneur/innovator role assumes in nowadays competitive organizations. Taking into account this reality, interdisciplinary perspectives to further our knowledge and understanding of the developments in entrepreneurship and organizational innovation areas are needed. Drawing on the latest developments and best practice, this book looks to examine the new advances in the subjects of organizational innovation, resultant from the last changes that are taking place and how they affect the management as well as the commitment and motivation of these organizations workers.

Following its main aims, this book looks to cover the field of *Entrepreneurship and Organizational Innovations* in six chapters. The first chapter presents "Mediating Effect of the HRM on the Relationship Between the SIMS and New Product Radicality"; the second chapter discusses "Reinventing Human Resource Management to Increase Organizational Efficacy". The third chapter contains information about "Employer Branding: Issues of Tailoring Your Message in the Modern Age", while the fourth chapter focuses "A Qualitative Investigation for Platform Model Conceptualization and Design: Propositions for a New Architecture". The fifth chapter deals with "The Utility of Human Resource Managers' Action: A Self-centred Perception by Different Organizational Actors", while the sixth chapter speaks about "About Competencies, Creativity and Innovation in the Portuguese Textile and Clothing Sector".

Looking to highlight some of the most recent developments in this research field, *Entrepreneurship and Organizational Innovation* can be seen as good support and tool to be used by academics, researchers, managers, engineers, practitioners and other professionals in related matters with management and business innovation.

The editors acknowledge their gratitude to Springer for this opportunity and for their professional support. Finally, we would like to thank all chapter authors for their interest and availability to work on this project.

Braga, Portugal — Carolina Machado
Aveiro, Portugal — J. Paulo Davim

Contents

About the Editors

Carolina Machado received her Ph.D. degree in Management Sciences (Organizational and Politics Management area/Human Resources Management) from the University of Minho in 1999, Master degree in Management (Strategic Human Resource Management) from Technical University of Lisbon in 1994 and Degree in Business Administration from University of Minho in 1989. Teaching in the Human Resources Management subjects since 1989 at University of Minho, she is since 2004 Associated Professor, with experience and research interest areas in the field of Human Resource Management, International Human Resource Management, Human Resource Management in SMEs, Training and Development, Emotional Intelligence, Management Change, Knowledge Management and Management/HRM in the Digital Age. She is Head of the Department of Management and Head of the Human Resources Management Work Group at University of Minho, as well as Chief Editor of the International Journal of Applied Management Sciences and Engineering (IJAMSE), Guest Editor of journals, Books Editor and Book Series Editor, as well as reviewer in different international prestigious journals. In addition, she has also published both as editor/co-editor and as author/co-author several books, book chapters and articles in journals and conferences.
email: carolina@eeg.uminho.pt

J. Paulo Davim received his Ph.D. degree in Mechanical Engineering in 1997, M.Sc. degree in Mechanical Engineering (materials and manufacturing processes) in 1991, Mechanical Engineering degree (5 years) in 1986, from the University of Porto (FEUP), the Aggregate title (Full Habilitation) from the University of Coimbra in 2005 and the D.Sc. from London Metropolitan University in 2013. He is Senior Chartered Engineer by the Portuguese Institution of Engineers with an MBA and Specialist title in Engineering and Industrial Management. He is also Eur Ing by FEANI, Brussels, and Fellow (FIET) by IET, London. Currently, he is Professor at the Department of Mechanical Engineering of the University of Aveiro, Portugal. He has more than 30 years of teaching and research experience in Manufacturing, Materials, Mechanical and Industrial Engineering, with special emphasis in Machining & Tribology. He has also interest in Management,

Engineering Education and Higher Education for Sustainability. He has guided large numbers of postdoc, Ph.D. and master's students as well as has coordinated and participated in several financed research projects. He has received several scientific awards. He has worked as an evaluator of projects for European Research Council (ERC) and other international research agencies as well as examiner of Ph.D. thesis for many universities in different countries. He is the Editor in Chief of several international journals, Guest Editor of journals, Books Editor, Book Series Editor and Scientific Advisory for many international journals and conferences. Presently, he is an Editorial Board Member of 30 international journals and acts as a reviewer for more than 100 prestigious Web of Science journals. In addition, he has also published as editor (and co-editor) more than 100 books and as author (and co-author) more than 10 books, 80 book chapters and 400 articles in journals and conferences (more than 200 articles in journals indexed in Web of Science Core Collection/h-index 51+/7500+ citations, SCOPUS/h-index 56+/10500+ citations, Google Scholar/h-index 71+/16500+).
email: pdavim@ua.pt

Mediating Effect of the HRM on the Relationship Between the SIMS and New Product Radicality

M. Martinez-Costa, D. Jimenez-Jimenez, Y. Castro-del-Rosario and Ledian Valle-Mestre

Abstract Launching new products to market has become a requirement for all companies seeking to survive in competitive markets. If they also intend to introduce more radical innovations, organizations should have more structured innovation management systems aimed at directing each element immersed in this process. As it could not be otherwise, the people involved in the innovation process play a key role. The purpose of this study is, first, to analyse the influence of a standardized innovation management system, based on the UNE 166002 standard, on the development of new radical products. Additionally, the mediating role of HR practices that explains how the innovation management system translates into radical products is analysed. The results of the study support our proposals and highlight the importance of personnel management in innovation.

Keywords HRM · Standardized Innovation Management System · UNE 166002 · New Product Radicality

1 Introduction

Globalization, high competitiveness and speed in the changes are characteristics of the environment that lead the company to the adoption of innovation as a tool for the creation and maintenance of sustainable competitive advantages.

Despite the broad conceptual debate regarding what the term "innovation" represents, it is generally assumed as a systemic process focused on the organized search for changes, as well as the analysis of the opportunities they can offer, under new organizational structures (Christensen et al. 2008; Schumpeter 1934, 1942). Hence, its management is one of the points of interest in the analysis of some of the studies developed in this field.

In order to develop these innovative activities, different studies and institutions have suggested creating a series of standards that facilitate the innovative work of

M. Martinez-Costa · D. Jimenez-Jimenez (✉) · Y. Castro-del-Rosario · L. Valle-Mestre
Departamento de Organización de Empresas y Finanzas, Universidad de Murcia, Murcia, Spain
e-mail: danieljj@um.es

C. Machado and J. P. Davim (eds.), *Entrepreneurship and Organizational Innovation*, Management and Industrial Engineering, https://doi.org/10.1007/978-3-030-19289-1_1

the company (Pellicer et al. 2008). In addition to improving internal processes for innovation, standardization will also help manage the large amount of documentation, information and knowledge that must be managed in relation to innovation and R&D. Therefore, the management of these activities demands a regulatory framework with guidelines and methodologies for innovation management (Mir-Mauri and Casadesus-Fa 2008). This has resulted in the creation of a standardized innovation management system (SIMS) as a comprehensive system that allows managing all innovation activities in the company. Although many studies have conducted research in other fields of standardization, there is a great lack of the effect of standards that promote innovation and the literature is limited to theoretical reviews and some case studies (Pellicer et al. 2012).

The management of innovation requires the efficient development of activities and the use of resources, generating and using new knowledge in the innovative process. It must also promote the generation of ideas for the creation or improvement of products, processes and services and, at the same time, it seeks to improve the transfer of these ideas to the production and marketing phases. At the same time, it is sought to enhance the transfer of these ideas to the production and marketing phases. Therefore, the R&D management structure that supports all the complex nature that it comprises and the interactions that are established is very important (Arzola et al. 2008).

Among many factors, the personnel of the company are a fundamental element, since it is responsible for generating and implementing the ideas that will lead to the innovations. Hence, both the SIMS and the human resource management are a priority for achieving the results of the innovative organization.

Based on the above, this chapter presents a study that analyses the relationship and effects of the standardized innovation management system, specifically the one proposed by the Spanish Standard UNE 166002: 2014, in the development of radical innovations, as well as the paper mediator of a human resource management system.

2 Theoretical Background

2.1 Radical Innovations

In the last decades, studies on innovation have increased. In this sense, researchers have focused on innovation as a key factor for the competitive advantage and survival of companies (Damanpour 1991; De Jong and Vermeulen 2006; Kleinschmidt and Cooper 1991; Subramanian and Nilakanta 1996; Damanpour et al. 1989). However, there is not always an agreement on the concept of innovation. Some authors refer to innovation as a process, while others focus on innovation as the result of a process. The definition offered by the OECD includes the key aspects that characterize the phenomenon of innovation and considers it as:

> ...as the implementation of a new or significantly improved product (good or service), or process, a new marketing method, or a new organisational method in business practices, workplace organisation or external relations. (OECD 2005)

The literature also refers to the different ways of innovating within the company pointing out various types of innovation, such as product/process, radical/incremental, technological/management, originating in the market and in relation to the impact economic. More or less differently, each of these types of innovation contributes to improving the company's processes and products with the aim of improving competitiveness.

However, innovation is a complex and very risky process that requires the identification of the determinants of the process. That is why innovation management models have acquired great importance. These models have been designed to provide guidelines that allow companies to generate competitive advantages in a systematic way and that, in turn, can respond to an environment of high complexity and competition.

Several authors point out that the management of innovation requires, among others, creativity and ideas, organization and people, R&D, the launch of new products, production and operations, technological strategy, marketing, the portfolio, projects, leadership, organizational culture, the establishment of cooperation networks and alliances, customer orientation, environmental monitoring, knowledge management, measurement and monitoring of results and people management (Goffin and Pfeiffer 1999; Goffin and Mitchell 2005; Dodgson 2000; Velasco Balmaseda and Zamanillo Elguezabal 2008). From this perspective, the management of innovation does not refer only to technological change or the launch of new products, but also deals with technology, people, culture, communication and organization, as well as processes of business. For Tidd et al. (1997), innovation constitutes the ability to recognize, align, acquire, generate, choose, execute, implement, learn and develop the organization. However, there is not always a terminological coincidence to name these skills, referring the literature that studies the SGI to components, key areas, critical areas, fundamental principles or basic criteria, or even adopting a more dynamic approach, essential processes, facilitating processes, basic phases or activities.

Finally, it has been observed that the type of innovation influences the type and quantity of efforts made in R&D or in the results of the company. Therefore, it is convenient to focus on a specific type of innovation and on the factors that power it. Of these types, undoubtedly, product innovation is the one that has greater relevance because it directly influences the company's results. According to the Oslo Manual, product innovation is understood as:

> the introduction of a good or service that is new or significantly improved with respect to its characteristics or intended uses. This includes significant improvements in technical specifications, components and materials, incorporated software, user friendliness or other functional characteristics. OECD (2005)

However, the definition of product innovation itself introduces the possibility of making a smaller or greater amount of changes in the new product, giving rise to the concept of radicality of innovation. Radical innovation must be understood as that which represents a novel and totally different contribution to what already existed and which, according to Afuah (1999: 20), for their exploitation they require a technological knowledge very different from the existing one. On the other hand, incremental innovations are those that start from prior knowledge and consist of substantial improvements of the product, service or process that, although they have a certain degree of novelty, do not clearly break with what existed up to that moment. Therefore, incremental innovation introduces relatively fewer changes in current products, exploiting design and often reinforcing the dominance of companies (Ettlie et al. 1984; Dewar and Dutton 1986; Henderson and Clark 1990: 9). As for the strategy of imitation, it is indicated that it will only try to incorporate some type of innovation that it considers profitable for your company. From there, it is a shared criterion in the studies developed in this field that the innovation strategy should lead to the development of incremental or radical innovations, to a greater extent than those derived from imitation.

Regarding radical innovation, some authors have found evidence that the most innovative products achieve greater acceptance in the market (Bayus et al. 2003), providing the company with a competitive advantage in those products (Hultink and Langerak 2002). In addition, and by definition, the radical products are substantially different from the rest, which means that the company can operate it alone, at least for a time, which will result in a greater margin of exploitation of it. For this reason, it is generally considered that the radicality of new products increases the success obtained with innovation (Deshpandé et al. 1993; Baldwin and Johnson 1996; Gatignon and Xuereb 1997; Xin et al. 2009).

According to the literature, just over 70% of radical ideas and more innovative projects are generated through formal planning activities (Barczak et al. 2009), compared to 60% for incremental projects. A simplified, efficient and agile approach quickly brings innovative products to the market (Cooper 2008).

The importance of the changes suggested by radical innovation requires that companies make greater efforts in innovation than the creation of incremental innovations would require. That is why, highly innovative companies have designed a series of internal processes that ensure that companies can develop successful projects of new products. Some of these processes have been gathered in standardised innovation management systems that try to contribute to the efficiency of this innovative process.

2.2 The UNE 166000 Standardized Innovation Management System

The interest in the area of innovation has promoted the creation of different systems that make it easier for companies and institutions to promote business innovation. For this reason, different initiatives have emerged in each country, such as in Denmark (2010 DS-hæfte 36:2010—User-oriented innovation management), France (2013 FD X50-271:2013—Innovation management—Guide for innovation management implementation), the UK (2008 BS 7000-1:2008—Design management systems—Part 1: Guide to managing innovation), Portugal (2007 NP4457:2007— R&D&I Management: Requirements of the R&D&I management system), Europe (2013 CEN-TS 16555-1:2013— Innovation management: Innovation management system) and internationally (2018 anticipated ISO 50501—Innovation management system guidelines) (Mir et al. 2016).

Similarly, in Spain, an innovation management system known as the UNE 166000 standard has also been created by the Spanish Standardization Agency (AENOR). The UNE 166000 family consists of several standards, three of which are certifiable (166001, 166002 and 166006), while the other standards help in the implementation of R&D management systems. The aim of the UNE 166002 standard is

> to provide guidelines that go beyond the requirements established in other management system standards, in order to consider both the effectiveness and efficiency of an R&D management system and, therefore, the potential to improve the results, as well as the improvement of the procedures for the internal transfer of these results to optimise the technological innovation processes of the organization. (AENOR 2014)

According to Navarro Cabeza (2005), this type of standardization facilitates organizations that decide to implement R&D activities and obtain advantages such as (1) the systematization of R&D activities to make better use of resources and know-how; (2) the promotion of R&D as a differentiating factor of competitiveness and corporate reputation; (3) the promotion of technology transfer and the facilitation of tax relief and (4) the improvement of the participation and motivation of employees, as well as the satisfaction of shareholders. In any case, this series of standards allows us to address R&D tasks and carry them out more efficiently, as well as to develop innovation in an excellent way. According to this standard, the organization must (AENOR 2014):

- Identify the R&D activities that should be the object of the R&D management system and apply them through the organization.
- Determine the sequence and interaction of these activities.
- Determine the criteria and methods necessary to ensure that both the operation and control of these activities are effective.
- Ensure availability of resources and information necessary to support the operation and monitoring of these activities.
- Carry out the monitoring, measurement and analysis of these activities and establish the procedures to carry them out.
- Implement the necessary actions to achieve the planned results and the continuous improvement of these activities.
- Establish and document protection mechanisms and exploitation of results.
- Define and document a process map that allows to visualize the main elements of the system, its sequence and the interrelationships between these elements.

The SIMS based on the UNE 166002: 2014 provides guidance for (a) understanding the context of the organization; (b) establishing the leadership and commitment of management; (c) planning the development of R&D; (d) identifying and promoting the factors that support R&D; (e) developing the process of R&D management.

In that sense, the technological orientation includes behaviours such as substantial investments in research and development, use of the latest cutting-edge technologies and sophistication in the new product development and proactive exploration, acquisition and rapid integration of new technologies inside and outside the industry (Gatignon and Xuereb 1997). Furthermore, it is expected that a company with a strong technological orientation will develop and incorporate unique and different ideas for new products based on superior technologies (Kim et al. 2013). This greater innovative capacity will allow the company to increase the chances of success of obtaining radically different products that are subject to more planned research and the incorporation of new technology.

This study focuses on this last point, and it suggests that the implementation of the standard can help companies to develop new radical products that are essential for the survival of companies. Thus:

H_1: The SIMS is positively related to radical innovation.

2.3 Mediating Effect of the HR System in the Management of Innovation

As it has been pointed out, the regulation establishes that the organization must continually improve the suitability and effectiveness of the system through the strategy and the R&D policy, leadership, objectives and planning, the processes that give support for R&D and performance evaluation. During the last decades, the literature has recognized the transcendence of people in the processes of product generation. Therefore, different authors have tried to study what type of HR practices stimulate the development of innovation in the company. Among others, Jiménez Jiménez and Sanz Valle (2008) identified a system of HR practices that favoured the development of innovation in the company. The idea behind these studies is that a certain HR management can contribute decisively to the stimulation of innovative behaviour in employees that allows generating ideas, adopting risks and solving the problems that arise in the process of new products.

Deepening in the literature, during the last decade, the literature has shown a growing interest in the role of the management of human resources to stimulate the results of innovation (Laursen and Foss 2003; De Saa-Perez and Diaz-Diaz 2010; Donate et al. 2016; Chowhan 2016; Lopez-Cabrales et al. 2009) in a general way and, especially, referring to the radicality of the product (Beugelsdijk 2008; De Winne and Sels 2010). The basic rationale is that if an organization intends to generate products radically different from what exists, it must be creative and, therefore, adopt practices that encourage the creativity of employees (Amabile et al. 1996). Consequently, the role of HR management in this context should be to provide the necessary conditions for employees to provide the most creative, surprising and unique ideas for the development of new products. This must be translated to all HR practices.

For example, a *job design* that increases autonomy through empowerment and variety through job rotation influences employees' motivation to be creative (Jiang et al. 2012). Several studies state that, in general, a team with a high degree of freedom, together with the ability to adapt processes, autonomy and empowerment, is appropriate for a project that implies a high degree of novelty (Yap and Souder 1994; Lewis et al. 2002). In addition, job rotation, (Jiang et al. 2012) flexible schedules and decentralized organizational structures were positively related to product innovations, especially those that are more radical (Beugelsdijk 2008).

In relation to the practice of *selection*, this plays an essential role in the detection of the behaviours necessary for the development of new original products. For example, it is necessary to select candidates with willingness to new experiences, in particular, frequently mentioned as a potential background for the creativity of employees (Williams 2004) and candidates with cognitive capacity (Hunter et al. 2012; Jiang et al. 2012). In addition, personal characteristics that can lead to creative design solutions are needed, for example, a consumer-oriented mind (Leonard and Rayport 1997), structured imagination skills (Ward 1994) and a flexible mentality (Amabile 1988). The selection should focus on candidates who have social competences. Thus, Pérez-Luño et al. (2011) found that social competences affect performance in exploration activities. These authors analysed social competence as an influential factor for success in radical product development projects. This is supported by Edvardsson and Oskarsson (2008), who state that social skills are important factors in hiring personnel in complex knowledge environments. That is to say, an intense and extensive search and a careful selection entails that the companies have a highly creative staff for the development of new original products.

In the same way, *training and development* also play a fundamental role in the development of new radical products. Licuanan et al. (2007) show that training that encourages creativity in teams can facilitate the generation of new ideas. By offering training opportunities that can increase the knowledge base of individuals or their relevant creativity skills, employees try to be more creative in their work (Shalley et al. 2004). In addition, radical innovation is based on intrinsically disruptive knowledge and driven by social capacity, the combination of new knowledge and a variety of perspectives (Beugelsdijk 2008). On the other hand, employees should receive guidance in establishing career plans that help them experience various job opportunities beyond the limits of a single experience in order to be innovative (Kang et al. 2007). These means are effective in promoting knowledge and evaluating new solutions (Zhou 1998).

In relation to *appraisals*, although there are some tests that relate it positively with innovation (Shipton et al. 2006), there is mixed evidence about their relationship as a background of creativity (Zhou and Shalley 2003; Jiang et al. 2012). Some researchers indicated that when employees wait for an evaluation, they may be less creative due to the anxiety associated with the evaluation (Byron et al. 2010). Others have connected the evaluation to higher levels of motivation and creativity (Egan 2005). Performance evaluation that takes into account knowledge and appropriate skills can encourage creativity among employees (Shipton et al. 2005). Risk-taking is another important task of performance evaluation, since innovation is always accompanied by risk (Gupta and Singhal 1993). Therefore, performance evaluation focused on innovation should not only consider visible results, but should also recognize the progress made and concentrate on quite subjective performance indicators, such as proactivity (Parker et al. 2006), creativity, the motivation and the assumption of risks (Gupta and Singhal 1993; Amabile 1998). In addition, for the development of new original products, evaluation activities should be designed with the objective of measuring other individual competencies and their potential, such as interpersonal skills that measure the ability to maintain positive affective relationships (Bissola et al. 2014). This is because creativity improves when people work together and observe the creative behaviours of their peers. Finally, for radical innovations, it may be more beneficial to leave a certain openness and ambiguity as to what results of knowledge are valued, in order to promote diversity and achieve a truly radical innovation (Andreeva et al. 2017).

On the other hand, in terms of *compensation*, so that the novelty is related to performance, employees must perceive a clear signal that they will be supported, instead of being sanctioned for their novelty. By providing extrinsic rewards for creativity, organizations are explicitly pointing out that creativity is a key outcome (Sue-Chan and Hempel 2016). Further, extrinsic reward serves to foster creativity (Dul et al. 2011; Shalley et al. 2004). When creativity is driven by personal and intrinsic motivations (Amabile 1983), there is no guarantee that the creativity of the employee is directed towards the organizational objectives. However, extrinsic rewards can help ensure the congruence of the individual and organizational objectives underlying creativity. In contrast to the above, Beugelsdijk (2008) showed that radical innovations benefit more from reactive efforts such as autonomy and empowerment, that this type of innovation could actually be impeded by monetary incentives and that, on the contrary, incremental innovations benefit greatly from training and pay based on performance.

These arguments show that a set of human resource practices oriented towards innovation (Fig. 1) leads to the creation of an adequate climate for the development of radical new products.

However, the question that arises is whether the management systems mentioned above can stimulate this system of HR practices as a mechanism to achieve radical innovations. In the case of innovations of the radical type, which imply that the resulting products differ significantly from those existing in the market, they should encourage the incorporation of new ideas and knowledge, with the aim of breaking with what exists up to now and creating a different concept (Kach et al. 2012). This will require that the SGI promotes such changes.

For this, the SIMS ensures that there is a high involvement of senior management in innovation, as this will in turn facilitate the motivation and involvement of employees. The lack of participation of the leader can be detrimental to the DNP projects, especially for innovations with a high degree of novelty, because it can decrease the effectiveness within the team and the communication of the team (Moenaert et al. 2000). Their involvement will facilitate the understanding of the objectives to the staff (Sundström and Zika-Viktorsson 2009) and will increase their responsibility (Andriopoulos and Lewis 2009), promoting the search for novel products. In

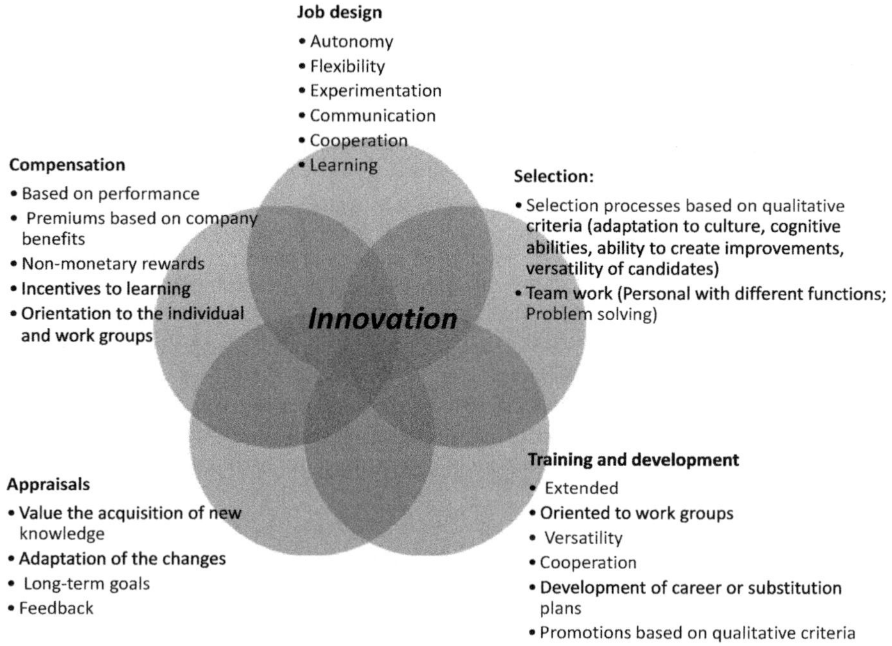

Fig. 1 HRM practices for innovation

addition, the direction promoted by this SIMS encourages employees to take risks, empowers staff and encourages employees to generate unique and innovative ideas for the development of new products in innovation teams (Amabile 1988; Amabile et al. 1996; Cummings and Oldham 1997).

Additionally, the standard indicates the importance of collaboration with other employees and external agents. Different studies have shown how the collaboration and commitment of the employees within the teams fosters the radicality of the new products (Qiu et al. 2009; Dahl and Moreau 2002). In this sense, team collaboration can have synergistic effects on innovation efforts and can be beneficial, mainly because it promotes the assumption of risks, a greater tolerance to errors (Caldwell and O'Reilly 2003; Cowan et al. 2007) and the success of new products (Kach et al. 2012).

Also, the creation of new knowledge is fundamental for the development of innovations in the SIMS. This involves the selection of new employees with new ideas or the learning of the employees of the company. In this sense, the SIMS points out the need to seek client learning (Gatignon and Xuereb 1997; Chuang et al. 2015; Kim et al. 2012) or other agents external to the company (Karkkainen and Ojanpera 2006; Bahemia and Squire 2010) with the aim of increasing the radicality of the new products. Therefore, the internal and external collaboration in R&D allows employees to learn new skills or acquire external knowledge, increasing their capacity and development, and promoting the development of highly innovative new products.

Table 1 HRM practices under UNE 166002:2014

Job design	Employee participation, flexible job design, autonomy in tasks, employee empowerment, communication and teamwork
Selection	Selective contracting aimed at innovative behaviours, social skills and the culture of the company, selection of candidates based on their ability to meet identified needs
Training and development	Continuous training, aimed at meeting the needs of employees, staff development plans aimed at fostering innovation (specific technical knowledge, creative skills, teamwork, problem-solving, decision analysis, risk/opportunity analysis and social skills), emphasis on the importance of R + D + i
Appraisals	Development of feedback, based on the performance of the team in promoting a favourable attitude to carry out initiatives and contributions of ideas
Compensation	Rewards based on the high performance of innovation, non-monetary incentives, incentive for original ideas

Finally, another aspect that concerns SIMS is the creation of a culture that supports innovation, developing performance evaluation systems, recognition of the work and offering incentives for the creation of successful ideas. In this sense, a reward system that recognizes creative work and the performance of activities related to innovation will promote the intrinsic motivation of a team, a main determinant of a creative climate (Krishnan and Ulrich 2001). Therefore, if an organization measures and rewards employees according to long-term results, customer relationships or the satisfaction of their needs, employees will be more intrinsically motivated to generate new ideas (Amabile 1988; Amabile et al. 1996; Im et al. 2013).

In consequence, different HRM practices are encouraged by UNE 166002 standard (see Table 1).

Although there are very few empirical studies, some works such as Terán (2009) or Ramis-Pujol (2005) confirm that in innovation management systems, HR management is a fundamental aspect when making decisions and establishing policies, with success in innovation.

Therefore, from the above, we propose the following hypothesis:

H_2: The system of HR practices oriented to innovation mediates the relationship between the SIMS and the radicality of the new product.

3 Research Methodology

3.1 Sample and Data Collection

The population used in this study includes Spanish manufacturing organizations, with more than 100 employees according to the SABI database (Analysis of Iberian

Balances), with at least 5 years old. The final population was 2.113 companies listed in the Spanish SABI database. The unit of analysis in this study is the company.

Data were collected through a telephone survey addressed to the director of R&D or new products and, otherwise, another company manager. They were given the choice of answering by phone, web page or email. The structured questionnaire was previously tested by academic and a small sample of managers.

A convenience sample of 200 firms was considered appropriate in order to test the research hypotheses. A total of 1.044 companies were randomly contacted in order to reach that number of valid questionnaires. It represents a response rate of 19.16%. A pretest was conducted in ten companies from different sectors to refine the questionnaire and check that all concepts are understood.

The survey was answered by top managers, mainly R&D managers (172). The sample was composed of medium-sized companies (the average size is 467 employees): 25.0% of the sample had fewer than 143 employees, 50% had fewer than 210 employees, and 75% had more than 350 employees. Moreover, the average number of years since its constitution was 37 years and the 91% of companies have international activity. Finally, on average, the companies in the sample showed the following characteristics: 116,808,488 € in operating incomes, 32,346,261 € on profits, 124,023,775 € in assets, 6.85% of return on assets and 1.56% of return of equity. In addition, chi-square distribution analyses revealed no significant differences between our sample and the population in terms of industry distribution, the number of employees and sales volume.

3.2 Measurements

Measures of the constructs were obtained from previous studies. In the research model, all the variables use five-point Likert scales (1 = "Strongly disagree"; 5 = "Strongly agree") for the perception of the respondent. The variables used are as follows:

Standardized system of innovation management has been computed following the methodology used in the application of the Innovation Management Standard 166002. We have processed this system as a formative construct, created from 15 indicators that compile the main requirements for an adequate Innovation Management. These indicators are derived directly from the standard 166002 of AENOR.

The HRM system has been created as a formative second-order factor construct made up of the 5 most important HR practices: job design, selection, training, performance appraisals and compensation. Each HRM practice has been considered as a formative construct with 5 items per practice from the studies of Lepak and Snell (2002) and Chen and Huang (2009).

Radical innovation has been measured with 5 items from Yang et al. (2014) adapted from previous studies (Gatignon et al. 2002). These measures have been treated as reflective.

The impact of common method bias was assessed using post hoc approaches. This potential problem was tested with a Harman factor test (Podsakoff and Organ 1986a). The results of the analysis of unrotated main components indicated that the variance of the common method did not represent a serious problem in our investigation because there were several factors with an eigenvalue greater than 2.

3.3 Analysis

Hypotheses were simultaneously tested using partial least squares (PLS), a structural equation modelling technique employing a principal component-based estimation approach (Chin 1998) using the SmartPLS 3 software (Ringle et al. 2005). PLS is a regression-based structural equation modelling (SEM) technique that makes no assumptions about data distributions. According to Podsakoff and Organ (1986b), PLS avoids many of the restrictive assumptions, underlying maximum likelihood techniques and ensuring against improper solutions and factor indeterminacy. In addition, PLS handles both reflective and formative constructions (Hair et al. 2006), as is the case of this research.

Using PLS involves following a two-stage approach (Barclay et al. 1995). The first step requires the assessment of the measurement model. This allows the relationships between the observable variables and theoretical concepts to be specified. This analysis is performed in relation to the attributes of individual items reliability, construct reliability, average variance extracted (AVE) and discriminant validity of the indicators of latent variables. In the second step, the structural model is evaluated.

To ensure an adequate quality in the measures used in the study, several criteria were evaluated. In the first place, and for the reflective scales, the reliability of the measurement scales was verified with the Cronbach alpha coefficient, obtaining in all cases a value higher than 0.7, considered adequate by the literature. The composite reliability ranged from 0.82 to 0.86, above the threshold of 0.7 (Nunnally 1978). An examination of the average variance extracted (AVE) revealed that all constructs exceeded the 0.50 cut-off set by the literature (Fornell and Larcker 1981). On the other hand, the R^2 value for the endogenous constructs exceeds the recommended minimum value of 0.1, which shows that the model developed is suitable for hypothesis testing (Table 2). The discriminant validity of the measures was then evaluated. As Fornell and Larcker (1981) suggested, the AVE for each construct should be greater than the correlations of squared latent factors between pairs of constructs. In addition, Table 2 also shows that the heterotrait–monotrait ratio (HTMT) values (elements above the diagonal) are below 0.85.

Consequently, it was determined that all the variables showed satisfactory discriminant validity. In short, our model has good convergent validity, reliability and discriminant validity.

Table 2 Reliability and validity of the measurement model

	Reliability			*Correlations*[a]		
	α	CR	AVE	SIMS	HR system	Prod. Radical
SIMS	0.954	0.960	0.650	**0.806**	0.621	0.373
HR system	0.771	0.843	0.519	0.555	**0.720**	0.445
Prod. radical	0.933	0.949	0.790	0.358	0.391	**0.889**

Note α = Cronbach's alpha; CR = composite reliability; AVE = average variance extracted
[a]Diagonal elements (bold figures) are the square root of the variance shared between the constructs and their measures. Below diagonal elements are the correlations between constructs. Above diagonal elements are the heterotrait–monotrait ratio (HTMT) values; $SRMR_{(Saturated\ Model)} = 0.067$; $SRMR_{(Estimated\ Model)} = 0.067$; d_ULS = 1.233; d_G = 0.714

4 Results

First, to confirm measurement model, several tests have been developed (Henseler et al. 2016). Specifically, these analyses (saturated model) provided standardized mean squared residual (SRMR = 0.067), unweighted minimum squares discrepancy (dULS = 1.233) and geodetic discrepancy (dG = 0.714) values within acceptable limits (values less than percentile levels based on bootstrap 95%). This implies that the model has an acceptable adjustment in PLS-SEM (Albort-Morant et al. 2018).

Second, to test our hypotheses, SmartPLS 3.2.8 has been used with the bootstrap resampling method (Chin 1998). Figure 2 shows the results of the proposed model. The structural model resulting from the PLS analysis is summarized in Fig. 2, where the explained variance of endogenous variables (R^2) and the standardized path coefficients (β) are shown.

The stability of the estimates is examined by using the *t*-statistics obtained from a bootstrap test with 5000 resamples. Table 3 sets out the model statistics, the path coefficients and the *t* values observed with the level of significance achieved from the bootstrap test.

First, the HRM system second-order construct explains all the personnel HRM practice system (job description: $\beta = 0.258$; selection: $\beta = 0.230$, training: $\beta = 0.251$; appraisal: $\beta = 0.263$; compensation: $\beta = 0.282$).

Second, we tested the effect of the effectiveness of an SIMS based on the UNE 166002 standard on the development of new radical products ($\beta = 0.555, p < 0.001$). This allows us to accept the first hypothesis and to understand that companies with a SIMS have better opportunities to develop radical innovations.

Our second hypothesis tests the mediator role of HRM system in order to explain the relationship between SIMS and radical innovations. Consequently, we have examined the effect of SIMS on HRM system ($\beta = 0.203$, $p < 0.01$) and between HRM system and product radicality ($\beta = 0.278$, $p < 0.001$). Both paths are significant. Furthermore, we have tested in an additional model without the HRM system the direct effect between SIMS and product radical ($\beta = 0.361, p < 0.001$). Comparing

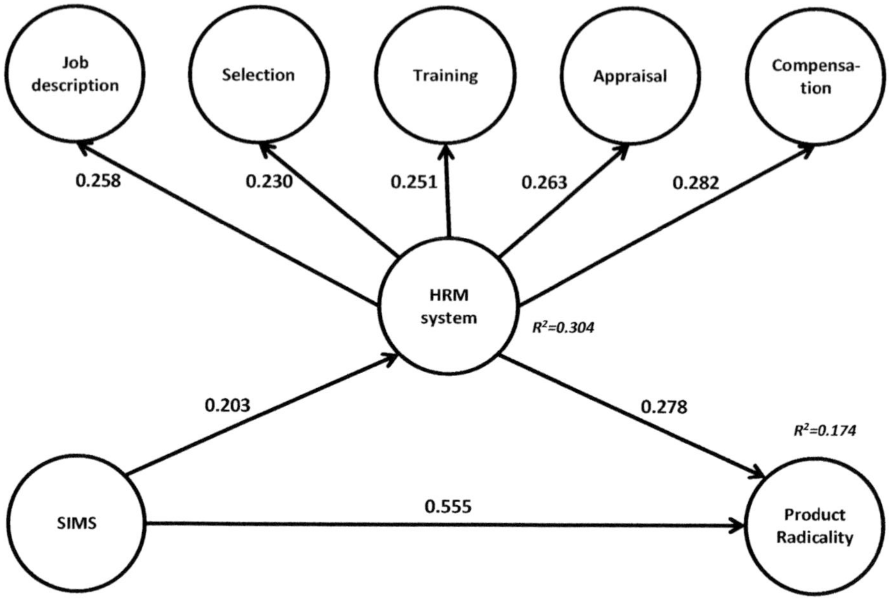

Fig. 2 Structural model

Table 3 Results of the proposed model

Paths	Coefficient	Std. Deviation	T value
Direct effect			
MIS → P. Radicality	0.203	0.081	2.509**
MIS → HRM system	0.555	0.047	11.855***
HRM system → P. Radicality	0.278	0.076	3.677***
Second order			
HRM system → Job description	0.258	0.024	10.575***
HRM system → Selection	0.230	0.027	8.408***
HRM system → Training	0.251	0.026	9.590***
HRM system → Appraisal	0.263	0.024	11.117***
HRM system → Compensation	0.282	0.023	12.150***
Total effect			
MIS → P. Radicality	0.358	0.066	5.431***

*$p < 0.1$, **$p < 0.05$, ***$p < 0.01$

with the model with mediation, we observed that effect of SIMS on product radicality is higher in the direct model than in the mediation model ($\beta = 0.263$, $p < 0.01$). Moreover, in Table 3, the total effect is also shown ($\beta = 0.358$, $p < 0.001$). These results suggest the existence of a partial mediation role of HRM system on the relationship between SIMS and radical innovation, what confirms hypothesis two.

5 Conclusions

Innovation is most of the time the only way for the company to survive in certain markets, given the imperative need to continue launching new products that attract the attention of new customers or meet the expectations of those loyal to the company. In any case, innovation is an essential mechanism to increase the company's results.

Although theoretically it is simple to foresee this relationship, the complexity that leads innovation management incorporates a series of unknowns into this equation. Fundamentally, the management of innovation requires an investment and certain costs in the development of new products, at the same time of time to generate certain routines that result in successful processes. In addition, the results of this process are always uncertain. In fact, the failure rates of new products are very high, since problems arise in the design, development or even marketing of innovation.

To mitigate these drawbacks, the literature has highlighted the need to design systems that regulate all R&D activities in a structured manner, with the aim of formalizing this activity within the company and not allowing improvization to carry out the necessary processes that can hinder the success of innovation.

This is even more relevant in radical innovations projects. In this case, the investment and the possibility of failure of radical innovation are substantially greater. Not only is it about doing something new, but it must be different and at the same time be competitive in the market. Obviously, if it is achieved, the performance of this type of disruptive innovations can be enormous, generating even a monopoly position that can be extended over time while being copied by the competition.

But to get a disruptive and successful idea, you need to look for adequate financing, effective techniques for generating ideas, adequate leadership, problem-solving mechanisms, market studies or an adequate commercialization of the new idea. Experience has shown that these processes must be structured to take them more effectively, under a system of innovation management in the company.

This has not been exempt from some controversy. The idea of standardizing innovation processes seemed not to work the same as in other management systems such as quality. From the perspective that standardization could kill the creativity and flexibility of innovative processes, some authors anticipated a negative relationship. However, the emergence of the standardization of innovation management did not require renouncing these mechanisms, but it had an impact on the need to structure activities to professionalize them and avoid failures in forecasting their processes.

In a similar way to other innovation management systems that have emerged in different countries, the Spanish regulation UNE 166002 allows organizations to be more innovative and promote the success of their innovations in products, services, processes, organizational designs or business models, thus contributing to the improvement of its results, its value and its competitiveness (AENOR 2014). In addition, this model is a general scheme that can be used as a point of reference by any organization, regardless of its size, in order to define or establish its own R&D process.

Among the different approaches highlighted by the standard, the most relevant focus on areas such as R&D management system and model; responsibility of top management; responsibility, authority and communication; management review; resource management; R&D activities; and measurement, analysis and improvement. All of them are necessary areas that must be properly developed and documented for their correct application.

Within resource management, the standard focuses on several aspects among which undoubtedly highlight human resources. Under this rubbish, their motivation, competence, awareness or training are key elements for this system to work. This is in line with what was argued by a great previous literature that suggests that personnel and HR practices are very relevant elements for the development of innovations in the company. In this line, the literature has stressed that innovation requires employees to develop ideas, be motivated, share information and solve problems. In short, they must show a favorable behavior to innovation (Sanz-Valle and Jiménez-Jiménez 2018). But this behaviour needs to be guided by a series of HR practices that show the attitudes demanded and paid for by the company.

HR practices cover different areas of personnel management such as job design, selection, training, appraisals or compensation. Although each of these areas focuses on certain aspects, an important key is that they are compatible with each other and that they seek the same objectives and obtain synergies in their application. It is for this reason the convenience of applying a system of consistent HR practices aimed at promoting innovation in the company.

In summary, the approach of this study states that in order to obtain radical innovations, companies must have innovation management systems that will enhance these results with different measures such as adequate personnel management.

The results obtained from a sample of manufacturing companies encourage these theses. In the first place, it has become clear that the application of SIMS has clear effects on the development of radical innovations. This confirms that this type of new products requires more structural processes that enhance the search for resources or the support of senior management as mechanisms for promoting and managing the innovative process.

In the second place, it has also been proven that the SIMS promotes the application of a series of HR practices favourable to innovation. In a special way, SIMS promotes the creation of a system of 25 practices that have been identified as determinants of innovation.

In the third place, it has been proven that said system of HR practices is related to the radicality of innovation. Although previous studies had suggested that HR management practices could enhance different types of innovation, there were some gaps in whether such practices could also promote the development of new products that were radically different from those that exist to date. The evidence corroborates these results and again highlights the importance of personnel management for innovation.

Finally, it has been verified whether there is mediation of such personnel practices in the relationship between the SIMS and the radicality of the innovation, answering the question implicit in the title of this study. The results show the existence of a partial mediation of HR management. This implies that part of the effect of the SIMS on the radicality of the new product is due to the fact that the standard favours the implementation of an adequate personnel management that results in employee behaviours that favour innovation.

In short, the results of our model show that the application of a SIMS will have a direct effect on the radicality of innovation and an indirect effect through personnel management practices.

5.1 Implications for Managers

The implications for managers that arise from this study are clear. In the first place, despite the risks of radical innovations, the literature highlights the innumerable advantages it provides for those companies that are competing in turbulent and dynamic environments.

However, for this and taking into account the risks of these innovations, the company must implement a management system that facilitates decision making and control of the entire innovation process. Our study has highlighted that the UNE 166002 standard has become an adequate SIMS for the promotion of this type of more disruptive innovations. This considerably facilitates the work of the managers of the companies by including in this guide the main actions that should be applied in the company to be successful in innovation.

Finally, company managers should not forget the management of the company's personnel. As shown, an adequate personnel management system is necessary for SIMS to provide the expected results. Therefore, from the human resource function a series of practices that affect the job design, selection, training, appraisal or compensation should be applied. These practices not only influence the obtaining of innovations, but also affect the behaviour of employees, guiding them towards greater creativity, risk-taking and innovation.

5.2 Limitations and Future Research Lines

The most relevant limitations of this study are based on the information collected. In the first place, this study has a cross-sectional sample that does not allow to observe effects that are delayed in time or to analyse in depth the causality of the relationships. Second, a single informant has been used. In this case, it would be appropriate to use different sources of information to get opinions from different executives of the company or even from the employees themselves. Finally, only information obtained from a survey is included, so it would be appropriate to incorporate another type of more objective data.

In addition to overcoming these limitations, future work could incorporate different levels of study with the multilevel methodology. In addition, variables such as employee behaviour, knowledge management or differentiating between different stages of innovation management should be incorporated. Finally, it would be appropriate to incorporate other innovation measures to identify the possibilities of success of the new product.

Acknowledgements The authors would like to thank the Ministerio de Economía, Industria y Competitividad of the Spanish Government for financing the research project ECO2017-88987-R (MINECO/FEDER;UE), co-financed from the European Union FEDER. Also, the authors gratefully acknowledge the financial support from Caja Murcia Foundation.

References

AENOR. (2014). *Gestión de la I+D+i: Requisitos del Sistema de Gestión de la I+D+i. UNE 166002*. Madrid: AENOR.

Afuah, A. (1999). *La dinámica de la innovación organizacional. El nuevo concepto para lograr ventajas competitivas y rentabilidad*. México: Oxford University Press.

Albort-Morant, G., Henseler, J., Cepeda-Carrión, G., & Leal-Rodríguez, A. L. (2018). Potential and realized absorptive capacity as complementary drivers of green product and process innovation performance. *Sustainability, 10*(2), 381–400.

Amabile, T. M. (1983). The social psychology of creativity: A componential conceptualization. *Journal of Personality and Social Psychology, 45*(2), 357.

Amabile, T. M. (1988). A model of creativity and innovation in organizations. *Research in Organizational Behavior, 10*(1), 123–167.

Amabile, T. M. (1998). How to kill creativity. *Harvard Business Review, 76*(5), 77–87.

Amabile, T. M., Conti, R., Coon, H., Lazenby, J., & Herron, M. (1996). Assessing the work environment for creativity. *Academy of Management Journal, 39*(5), 1154–1184.

Andreeva, T., Vanhala, M., Sergeeva, A., Ritala, P., & Kianto, A. (2017). When the fit between HR practices backfires: Exploring the interaction effects between rewards for and appraisal of knowledge behaviours on innovation. *Human Resource Management Journal, 27*(2), 209–227.

Andriopoulos, C., & Lewis, M. W. (2009). Exploitation-exploration tensions and organizational ambidexterity: Managing paradoxes of innovation. *Organization Science, 20*(4), 696–717.

Arzola, M., Salazar, M., & Pérez, E. (2008). Gestión de la innovación en las PYMIS de Ciudad Guayana. In *II Simposio Internacional de Ingeniería Industrial: Actualidad y Nuevas Tendencias 2008 V Jornadas de Productividad, Calidad e Innovación*. Universidad de Carabobo, Valencia, Venezuela.

Bahemia, H., & Squire, B. (2010). A contingent perspective of open innovation in new product development projects. *International Journal of Innovation Management, 14*(4), 603–627.

Baldwin, J. R., & Johnson, J. (1996). Business strategies in more- and less-innovative firms in Canada. *Research Policy, 25*(5), 785–804.

Barclay, D., Higgins, C., & Thompson, R. (1995). The partial least squares (PLS) approach to causal modelling: Personal computer adoption and use as an illustration. *Technology Studies, Special Issue on Research Methodology, 2*(2), 285–309.

Barczak, G., Griffin, A., & Kahn, K. B. (2009). Perspective: trends and drivers of success in NPD practices: results of the 2003 PDMA best practices study. *Journal of Product Innovation Management, 26*(1), 3–23.

Bayus, B. L., Erickson, G. & Jacobson, R. (2003). The financial rewards of new product introductions in the personal computer industry. *Management Science, 49*(2), 197–210.

Beugelsdijk, S. (2008). Strategic human resource practices and product innovation. *Organization Studies, 29*(6), 821–847.

Bissola, R., Imperatori, B., & Colonel, R. T. (2014). Enhancing the creative performance of new product teams: An organizational configurational approach. *Journal of Product Innovation Management, 31*(2), 375–391.

Byron, K., Khazanchi, S., & Nazarian, D. (2010). The relationship between stressors and creativity: A meta-analysis examining competing theoretical models. *Journal of Applied Psychology, 95*(1), 201.

Caldwell, D. F., & O'Reilly, C. A. (2003). The determinants of team-based innovation in organizations the role of social influence. *Small Group Research, 34*(4), 497–517.

Chen, C.-J., & Huang, J.-W. (2009). Strategic human resource practices and innovation performance—The mediating role of knowledge management capacity. *Journal of Business Research, 62*(1), 104–114.

Chin, W. W. (1998). The partial least squares approach to structural equation modeling. In G. A. Marcoulides (Ed.), *Methodology for business and management. Modern methods for business research* (295–336). Mahwah, NJ: Lawrence Erlbaum Associates.

Christensen, C. M., Horn, M. B., & Johnson, C. W. (2008). *Disrupting class : How disruptive innovation will change the way the world learns*. New York: McGraw-Hill.

Chowhan, J. (2016). Unpacking the black box: Understanding the relationship between strategy, HRM practices, innovation and organizational performance. *Human Resource Management Journal, 26*(2), 112–133.

Chuang, F.-M., Morgan, R. E., & Robson, M. J. (2015). Customer and competitor insights, new product development competence, and new product creativity: Differential, integrative, and substitution effects. *Journal of Product Innovation Management, 32*(2), 175–182.

Cooper, R. G. (2008). Perspective: The Stage-Gate-Idea-to-Launch Process—Update, What's New, and NexGen Systems. *Journal of Product Innovation Management, 25*(3), 213–232.

Cowan, R., Jonard, N., & Zimmermann, J.-B. (2007). Bilateral collaboration and the emergence of innovation networks. *Management Science, 53*(7), 1051–1067.

Cummings, A., & Oldham, G. R. (1997). Enhancing creativity: Managing work contexts for the high potential employee. *California Management Review, 40*(1), 22.

Dahl, D. W., & Moreau, P. (2002). The influence and value of analogical thinking during new product ideation. *Journal of Marketing Research, 39*(1), 47–60.

Damanpour, F. (1991). Organizational innovation: A meta-analysis of effects of determinants and moderators. *Academy of Management Journal, 34*(3), 555–590.

Damanpour, F., Szabat, K. A., & Evan, W. M. (1989). The relationship between types of innovation and organizational performance. *Journal of Management Studies, 26*(6), 587–602.

De Jong, J. P. J., & Vermeulen, P. A. M. (2006). Determinants of product innovation in small firms a comparison across industries. *International Small Business Journal, 24*(6), 587–609.

De Saa-Perez, P., & Diaz-Diaz, N. L. (2010). Human resource management and innovation in the Canary Islands: An ultra-peripheral region of the European Union. *The International Journal of Human, Resource Management, 21*(10), 1649–1666.

De Winne, S., & Sels, L. (2010). Interrelationships between human capital, HRM and innovation in Belgian start-ups aiming at an innovation strategy. *The International Journal of Human Resource Management, 21*(11), 1863–1883.

Dewar, R. D., & Dutton, J. E. (1986). The adoption of radical and incremental innovations: An empirical analysis. *Management Science, 32*(11), 1422–1433.

Deshpandé, R., Farley, J. U., & Webster, F. E. (1993). Corporate culture. Customer orientation, and innovativeness in Japanese firms: A quadrad analysis. *Journal of Marketing, 57*(1), 23–37.

Dodgson, M. (2000). *The management of technological innovation: An International and Startegic Approach*. Oxford: Oxford University Press.
Donate, M. J., Peña, I., & Sánchez de Pablo, J. D. (2016). HRM practices for human and social capital development: effects on innovation capabilities. *The International Journal of Human Resource Management, 27*(9), 928–953.
Dul, J., Ceylan, C., & Jaspers, F. (2011). Knowledge workers' creativity and the role of the physical work environment. *Human Resource Management, 50*(6), 715–734.
Edvardsson, I. R., & Oskarsson, G. (2008). Knowledge management and value creation in service firms. *Measuring Business Excellence, 15*(4), 7–15.
Egan, T. M. (2005). Factors influencing individual creativity in the workplace: An examination of quantitative empirical research. *Advances in Developing Human Resources, 7*(2), 160–181.
Ettlie, J. E., Bridges, W. P., & O´keefe, R. D. (1984). Organization strategy and structural differences for radical versus incremental innovation. *Management Science, 30*(6), 682–695.
Fornell, C., & Larcker, D. F. (1981, February). Evaluating structural equation models with unobservable variables and measurement error. *Journal of Marketing Research, XXVII,* 39–50.
Gatignon, H., & Xuereb, J.-M. (1997). Strategic orientation of the firm and new product performance. *Journal of Marketing Research, 34*(1), 77–90.
Gatignon, H., Tushman, M. L., Smith, W., & Anderson, P. (2002). A structural approach to assessing innovation: Construct development of innovation locus, type, and characteristics. *Management Science, 48*(9), 1103–1122.
Goffin, K., & Mitchell, R. (2005). *Innovation management: Strategy and implementation using the pentathlon framework*. Basingstoke: Palgrave Macmillan Houndmills.
Goffin, K., & Pfeiffer, R. (1999). *Innovation management in UK and German manufacturing companies*. London: Anglo-German Foundation for the Study of Industrial Society.
Gupta, A. K., & Singhal, A. (1993). Managing human resources for innovation and creativity. *Research Technology Management, 36*(3), 41.
Hair, J. F., Anderson, R. L., & Tatham, W. C. (2006). *Multivariate data analysis*. Upper Saddle River, NJ: Pearson.
Henderson, R. M., & Clark, K. B. (1990). Architectural innovation: The reconfiguration of existing product technologies and the failure of established firms. *Administrative Science Quarterly, 35,* 9–30.
Henseler, J., Hubona, G., & Ray, P. A. (2016). Using PLS path modeling in new technology research: Updated guidelines. *Industrial Management & Data Systems, 116*(1), 2–20.
Hultink, E. J., & Langerak, F. (2002). Launch decisions and competitive reactions : An exploratory market signaling study. *Journal of Product Innovation Management, 19*(3), 199–212.
Hunter, S. T., Cushenbery, L., & Friedrich, T. (2012). Hiring an innovative workforce: A necessary yet uniquely challenging endeavor. *Human Resource Management Review, 22*(4), 303–322.
Im, S., Montoya, M. M., & Workman, J. P. (2013). Antecedents and consequences of creativity in product innovation teams. *Journal of Product Innovation Management, 30*(1), 170–185.
Jiang, J., Wang, S., & Zhao, S. (2012). Does HRM facilitate employee creativity and organizational innovation? A study of Chinese firms. *The International Journal of Human Resource Management, 23*(19), 4025–4047.
Jiménez Jiménez, D., & Sanz Valle, R. (2008). Could HRM support organizational innovation? *The International Journal of Human Resource Management, 19*(7), 1208–1221.
Kach, A., Azadegan, A., & Dooley, K. J. (2012). Analyzing the successful development of a high-novelty innovation project under a time-pressured schedule. *R&D Management, 42*(5), 377–400.
Kang, S.-C., Morris, S. S., & Snell, S. A. (2007). Relational archetypes, organizational learning, and value creation: Extending the human resource architecture. *Academy of Management Review, 32*(1), 236–256.
Karkkainen, P., & Ojanpera, T. (2006). Pushing technology to where the market will be: The case of Nokia. *International Journal of Technology Management, 34*(3–4), 203–216.

Kim, N., Im, S., & Slater, S. F. (2013). Impact of knowledge type and strategic orientation on new product creativity and advantage in high-technology firms. *Journal of Product Innovation Management, 30*(1), 136–153.

Kim, W., Di Benedetto, C. A., & Hunt, J. M. (2012). Consumer innovativeness and consideration set as antecedents of the consumer decision process for highly globalized new products: A three-country empirical study. *Journal of Global Scholars of Marketing Science, 22*(1), 1–23.

Kleinschmidt, E. J., & Cooper, R. G. (1991). The impact of product innovativeness on performance. *Journal of Product Innovation Management, 8*(4), 240–251.

Krishnan, V., & Ulrich, K. T. (2001). Product development decisions: A review of the literature. *Management Science, 47*(1), 1–21.

Laursen, K., & Foss, N. (2003). New HRM practices, complementarities, and the impact on innovation performance. *Cambridge Journal of Economics, 27*(2), 243–263.

Leonard, D., & Rayport, J. F. (1997). Spark innovation through empathic design. *Harvard Business Review, 75*(6), 102–115.

Lepak, D. P., & Snell, S. A. (2002). Examining the human resource architecture: The relationships among human capital, employment, and human resource configurations. *Journal of Management, 28*(4), 517–543.

Lewis, M. W., Welsh, M. A., Dehler, G. E., & Green, S. G. (2002). Product development tensions: Exploring contrasting styles of project management. *Academy of Management Journal, 45*(3), 546–564.

Licuanan, B. F., Dailey, L. R., & Mumford, M. D. (2007). Idea evaluation: Error in evaluating highly original ideas. *The Journal of Creative Behavior, 41*(1), 1–27.

Lopez-Cabrales, A., Pérez-Luño, A., & Valle-Cabrera, R. (2009). Knowledge as a mediator between HRM practices and innovative activity. *Human Resource Management, 48*(4), 485–503.

Mir, M., Casadesús, M., & Petnji, L. H. (2016, July–September). The impact of standardized innovation management systems on innovation capability and business performance: An empirical study. *Journal of Engineering and Technology Management, 41*, 26–44.

Mir-Mauri, M., & Casadesus-Fa, M. (2008). UNE 166002:2006: Estandarizar y sistematizar la I+D+i. La norma y la importancia de las TIC en su implementación. *DYNA, 83*(6), 325–331.

Moenaert, R. K., Caeldries, F., Lievens, A., & Wauters, E. (2000). Communication flows in international product innovation teams. *Journal of Product Innovation Management, 17*(5), 360–377.

Navarro Cabeza, I. M. (2005). Normalización y certificación de proyectos y sistemas de gestión de la I+D+I. *Técnica Industrial, 256*(March), 38–45.

Nunnally, J. C. (1978). *Psychometric theory*. New York: McGraw-Hill.

OECD. (2005). *Oslo manual. The measurement of scientific and technological activities. Proposed guidelines for collecting and interpreting technological innovation data* (2nd ed.). Paris: OECD/European Commission EUROSTAT.

Parker, S. K., Williams, H. M., & Turner, N. (2006). Modeling the antecedents of proactive behavior at work. *Journal of Applied Psychology, 91*(3), 636–652.

Pellicer, E., Correa, C. L., Yepes, V., & Alarcón, L. F. (2012). Organizational improvement through standardization of the innovation process in construction firms. *Engineering Management Journal, 24*(2), 40–53.

Pellicer, E., Yepes, V., & Martínez, G. (2008). Enhancing R&D&i through standardization and certification: The case of the Spanish construction industry. *Revista ingeniería de construcción, 23*(2), 112–121.

Pérez-Luño, A., Medina, C. C., Lavado, A. C., & Rodríguez, G. C. (2011). How social capital and knowledge affect innovation. *Journal of Business Research, 64*(12), 1369–1376.

Podsakoff, P. M., & Organ, D. (1986a). Reports in organizational research: Problems and prospects. *Journal of Management, 12*(4), 531–545.

Podsakoff, P. M., & Organ, D. W. (1986b). Self-reports in organizational research: Problems and prospects. *Journal of Management, 12*(4), 69–82.

Qiu, T., Qualls, W., Bohlmann, J., & Rupp, D. E. (2009). The effect of interactional fairness on the performance of cross-functional product development teams: A multilevel mediated model. *Journal of Product Innovation Management, 26*(2), 173–187.
Ramis-Pujol, J. (2005). *Guía práctica de la innovación para PYMES*. Palma de Mallorca Anetcom.
Ringle, C. M., Wende, S., & Will, A. (2005). *SmartPLS 2.0 (M3) beta*. Hamburg, Germany.
Sanz-Valle, R., & Jiménez-Jiménez, D. (2018). HRM and product innovation: Does innovative work behaviour mediate that relationship? *Management Decision* (in press).
Schumpeter, J. A. (1934). *The theory of economic development*. Cambridge: Harvard University.
Schumpeter, J. A. (1942). *Capitalism, socialism and democracy*. New York: Harper and Row.
Shalley, C. E., Zhou, J., & Oldham, G. R. (2004). The effects of personal and contextual characteristics on creativity: Where should we go from here? *Journal of Management, 30*(6), 933–958.
Shipton, H., Fay, D., West, M., Patterson, M., & Birdi, K. (2005). Managing people to promote innovation. *Creativity and Innovation Management, 14*(2), 118–128.
Shipton, H., West, M. A., Dawson, J., Birdi, K., & Patterson, M. (2006). HRM as a predictor of innovation. *Human Resource Management Journal, 16*(1), 3–27.
Subramanian, A., & Nilakanta, S. (1996). Organizational innovativeness: Exploring the relationship between organizational determinants of innovation, types of innovations, and measures of organizational performance. *Omega, 24*(6), 631–647.
Sue-Chan, C., & Hempel, P. S. (2016). The creativity-performance relationship: How rewarding creativity moderates the expression of creativity. *Human Resource Management, 55*(4), 637–653.
Sundström, P., & Zika-Viktorsson, A. (2009). Organizing for innovation in a product development project: Combining innovative and result oriented ways of working—A case study. *International Journal of Project Management, 27*(8), 745–753.
Terán, A. (2009). Desarrollo de un modelo de aplicación de la norma UNE 166002 sobre implantación de sistemas de innovación en las PYMIS venezolanas. Paper read at 10th Latin American and Caribbean Conference for Engineering and Technology, at Madrid.
Tidd, J., Bessant, J., & Pavitt, K. (1997). *Managing innovation: Integrating technological, market and organizational change*. New York: Wiley.
Velasco Balmaseda, E., & Zamanillo Elguezabal, I. (2008). Evolución de las propuestas sobre el proceso de innovación:¿Qué se puede concluir de su estudio? *Investigaciones Europeas de Dirección y Economía de la Empresa, 14*(2), 127–138.
Ward, T. B. (1994). Structured imagination: The role of category structure in exemplar generation. *Cognitive Psychology, 27*(1), 1–40.
Williams, S. D. (2004). Personality, attitude, and leader influences on divergent thinking and creativity in organizations. *European Journal of Innovation Management, 7*(3), 187–204.
Xin, J. Y., Yeung, A. C. L., & Cheng, T. C. E. (2009). Radical innovations in new product development and their financial performance implications: An event study of US manufacturing firms. *Operations Management Research, 1*(2), 19–28.
Yang, K.-P., Chou, C., & Chiu, Y.-J. (2014). How unlearning affects radical innovation: The dynamics of social capital and slack resources. *Technological Forecasting and Social Change, 87*, 152–163.
Yap, C. M., & Souder, W. E. (1994). Factors influencing new product success and failure in small entrepreneurial high-technology electronics firms. *Journal of Product Innovation Management, 11*(5), 418–432.
Zhou, J. (1998). Feedback valence, feedback style, task autonomy, and achievement orientation: Interactive effects on creative performance. *Journal of Applied Psychology, 83*(2), 261.
Zhou, J., & Shalley, C. E. (2003). Research on employee creativity: A critical review and directions for future research. In *Research in Personnel and Human Resources Management* (pp. 165–217). Emerald Group Publishing Limited.

Reinventing Human Resource Management to Increase Organizational Efficacy

José Rebelo dos Santos and Lurdes Pedro

Abstract This chapter focuses on human management and innovation. We aim to identify ongoing changes in innovative human resource practices and the reasons to adopt them. Specifically, we seek to (i) characterize the evolution of traditional to new practices; (ii) characterize each new practices and reasons to adopt it; (iii) quantify (or quality) the results of the new practices; (iv) identify the profiles of the organizations currently adopting these practices; and (v) present innovative practices of a private and a public organization. After a brief thematic literature review, our methodology goes over the analysis of empirical documents and performs semi-structured interviews to show what is doing in a private and in a public organization. The results show that in each organization that is implementing measures that value and care about people, there is recognition and belief that people valued by their employing organizations are more productive and generate better results.

Keywords Innovation · Human resources management · Best practices

1 Introduction

Human resource management assumes an increasingly important role in what comes to managing organizations.

The spotlight going to human resource management is justifiable since it is a resource that tends to be rare (i.e. is not available in abundance); a resource which is valuable (i.e. contributes to achievement of organizational goals), imperfectly imitable (i.e. cannot be reproduced by other organizations) and hardly substitutable (i.e. cannot be replaced by another form of resource) and as such should be managed well and very carefully.

As a result, the policies and practices of human resource management, especially the ones associated with innovation and organizational performance, can make a

J. R. dos Santos (✉) · L. Pedro
College of Business Administration, Polytechnic Institute of Setubal, Setúbal, Portugal
e-mail: jose.rebelo@esce.ips.pt

C. Machado and J. P. Davim (eds.), *Entrepreneurship and Organizational Innovation*, Management and Industrial Engineering, https://doi.org/10.1007/978-3-030-19289-1_2

difference and are determinant to the growing of the role of human resource management in organization management.

New or newly transformed HRM practices are emerging both in private organizations and in public—and thus more regulated—institutions.

The respect for each person's individuality and their personal specificities is underlying these new practices of human management which promote: (i) conciliation with family life; (ii) flexibility; (iii) new models of communication grounded in complex and sophisticated platforms to communicate; (iv) the delegation of responsibilities and the involvement of employees in decision making; (v) specific training, career management, mentoring and giving feedback to employees; (vi) and involvement as well as co-responsibilization.

Adopting these practices is not a trend, rather, it stems from a necessity of the current times. The traditional models of human resource management no longer respond as well in what comes to organizational efficacy and the new models oriented towards commitment and involvement (Lawler et al. 2010) have been presenting better and more visible results.

The central theme of this chapter is the management of people and innovation. This chapter will identify the occurring changes in what comes to innovative practices in human resources and why to adopt such practices. Specifically, this chapter will: (i) characterize from an evolutionary standpoint the traditional and the new practices; (ii) characterize each new practice and the reasons why it should be adopted; (iii) quantify (or qualify) the results of the new practices; (iv) identify the profile or profiles of organizations that are adopting the new practices; and (v) present the innovative practices of a private and a public organization.

In what comes to methodology, after a brief thematic literature review, we will analyse documents from empirical studies and perform semi-structured interviews to track what is being done in some organizations.

2 Theoretical Framework

In this section, we first discuss the importance of human resource management to then stress the strategic management of human resources. After this, we will reflect on innovation and creativity and its relation with human management models. Finally, we will identify emerging practices that integrate innovative aspects.

2.1 Human Resource Management and Strategic Management of Human Resources

Human resource management regards measures and activities that imply human resources and seek to maximize the efficiency and performance of individuals within organizations (Alis et al. 2012).

Traditionally, human resource management focused on short-term quantitative aspects regarding work hours, remunerations, social benefits and health and safety (Alis et al. 2012).

Currently, qualitative aspects that contribute to motivation such as the appreciation of people, opportunities for development and professional evolution, autonomy and communication have all gained importance (Alis et al. 2012).

Since the late eighties that the function of HR has come to be considered as a strategic role connected to the functioning of organizations, standing as a strategic partner and agent of change (Ulrich and Brockbank 2005).

The role of human resource management as now stopped being exclusively administrative and is also strategically connected to the different organizational dimensions (Melo and Machado 2015).

Within the present-day economy, lays in the consensually accepted assumption that it is a strategic management considered as (i) a shared function within the organization; in which (ii) HR are the main source of sustainable competitive advantage of the organization; and (iii) in addition, having an integrated approach of HRM with the remaining subsystems of management.

This implies that the strategies of human resource should be coherent and fully integrated within the company's strategies (vertical alignment) and coherent and fully integrated with the different functioning systems of HRM (horizontal alignment); both of which cannot contradict each other in their goals and in the effects they produce (Combs et al. 2006; Mabey and Salaman 1995).

Strategic human resource management corresponds to a "pattern of planned human resource deployments and activities intended to enable an organization to achieve its goals" (Wright and McMahan 1992: 298).

Human resource management corresponds to the structuring of operational activities in what comes to recruiting and selecting performance evaluation, training management and management of remuneration and mobilization of people (Dietrich and Pigeyre 2016). These practices have been studied within a theoretical framework or empirical as isolated and singular practices as well as within models that are made up of several human resource management practices (Pedro 2015).

We can consider core, or more traditional, functions of HRM, planning of human resources, analysis and description of functions, recruitment and selection, welcoming and integration, training, performance evaluation, remuneration and incentive management, competency management as well as internationalization and international management of human resources and expatriation (Melo and Machado 2015).

Management practices and practices of human resource management tend to renovate to allow for organizational choices to come to fruition in what comes to human resources (Dietrich and Pigeyre 2016).

2.2 *Creativity and Innovation*

Creativity must be present prior to any innovation. Creativity can be defined as conceiving new ideas (Amabile et al. 1996). For these authors, innovation may only take place if these ideas can be channelled into concrete action, and thus, we can say creativity is the starting point for innovation (Amabile et al. 1996). Innovation means putting into practice a new idea that works and is accepted (Barbieri 2012).

Organizational innovation entails planned changes in regard to the activity of the organization, implying something new and having as a goal to increase performance (OECD 2005).

Innovation can be radical or incremental according to whether it is aiming for something genuinely new which means substantial innovations or changes to what is already there which means lower intensity innovations (Curado 2018).

There are usually four types of innovation to be considered: organizational, product, process and marketing (OECD 2005).

Organizational innovation is unique and stems from a company's attempt to develop which can happen in any economic sector including public services such as health and education (OECD 2005).

Organizational innovation more than any of the others assumes an active process that is developed within the context of an independent structure (Sousa and Monteiro 2017).

Although organizational factors that may facilitate innovations include technologies, managerial support, specific firm design, the existing literature emphasizes the role of human capital (skills and capabilities of individuals) as the primary resource for innovation (Barney and Wright 1998) and it is clear that a firm's ability to produce new products or services is strongly linked to its human capital. As such, one of the main obstacles to innovation is the lack of qualified human resources and with lack of competencies (Baranano 2005).

Other researchers have highlighted the importance of the organizational climate for creativity and consequently for innovation to take place. Rodrigues and Veloso (2013) demonstrate that this relational dimension positively influences resource exchange and the coordination amongst the people involved, which, at the same time, creates value for the firm through its effect on innovation.

2.2.1 Human Resource Management Practices and Innovation

Innovative organizations hold good human resource management practices which make them be referenced in the ranking of best companies to work.

As a result, they are able to attract, contract and keep the best, which in turn comes with an increase in innovation and productivity.

Empirical studies show that recruitment and selection human resource management policies and practices can contribute to an increase in innovation and together with training, career and remuneration management and can ultimately promote innovative performance (Martins 2016).

It is also worth noting that in several firms, the size of the organization, its market, its sector as well as the socio-economic context can condition to what extent each practice produces results in promoting innovation (Amaral 2015).

Some organizational structure characteristics stimulate informality and the permanent redefinition of methods, tasks, functions, responsibility and authority through interaction amongst everyone (Baranano 2005).

According to recent studies (e.g. Lukkap 2014), new HRM practices grounded on maximizing work flexibility are emerging, both adapted to the life cycle of workers and responding to their needs and according to the life cycle phases. Human resource departments tend to manage employees' life cycle and develop reward and work condition systems that adjusted to each moment in life, encouraging trust and long-term relationships. New compensation programmes can be designed according to workers' needs (such as private pension plans, health insurance) and adapt to the workers' moments in life (for example, in case of wedding, disease or childbirth); they aim to not only maintain and keep workers satisfied but also guarantee individualized working relationships. It is from within this framework that a new laboural relationship model is designed, allowing for more room for negotiation amongst "employer—client", with more autonomy, initiative, lowering subordination levels, focused on valuing of competencies and the added value of work. These new approaches make it so human resource management focuses on identifying each worker's need, as individuals with professional expectations, valuing more personalized management. This approach opposes the traditional perspective which is oriented towards the fulfilment of needs of groups and work teams which brush over an individual's identity.

In what comes to recruitment and selection, processes that are focused on attracting the best allow the contract of a specialized professional with essential competencies in technological innovation processes (Baranano 2005).

New tools are being used today to attract and identify the best (for example, social networks, localization and streaming). Other tools, in what comes to new technologies, can also help cap those talents (e.g. storytelling, Tripad). Several companies provide their workers with relevant content and a webspace where they can have a blog focused on their interests, attracting external talent and reinforcing their reputation as a branded employer. With the same objective, other organizations develop referral programmes, in which their workers present to the company friends or acquaintances for job openings. In what comes to competency seeking, the tendency is to quickly connect workers and projects, with social media already allowing for monitoring of competencies, interests and motivations, within and outside the organization. Whilst in the recent past, companies might have sought candidates with years of experience through their recruitment process, now they more commonly look for candidates from different sectors but with the relevant competencies so that they innovate and are not stranded by the common practices of the sector. Employees enjoy challenges increasingly more so it is more common to see selection processes that test worker's competencies, making them face real work problems from the position they are applying to (Lukkap 2014).

With regard to vocational training, the programs implemented can increase motivation, since the most qualified human resources usually recognize training as a tool for professional development and personal development (Baranano 2005).

With new technologies, organizations tend to give workers more autonomy in the development of their work, making them responsible for their own professional development. Career planning comes as a responsibility in which the worker assumes a decisive role regarding which options to choose.

The company provides the necessary information so their decision is the best for them: presenting alternatives, comparisons, employability level and progress in comparison in between others and themselves, amongst other kinds of information.

Internal communication is mostly informal and goes in every direction which can be explained partly if the organizations are small (Baranano 2005), as is the case of many of these organizations. Entertainment through gamification is becoming easier and universal way of changing employees' habits and behaviours as well as communicating in organizations.

2.3 Employee Centric

Approaches to organizational change tend to emphasize either agents of change and the ways they conduct processes of change or the agents receiving the change (employees at large) (Armenakis and Harris 2009). The latter case is considered employee centric.

The company's stated philosophy is to be centred in the employee—taking care of their employees first. In turn, the company believes the employees will take care of their clients (Nohria et al. 2008).

If the company does everything for their employees to be happy they assume that in turn, their employees will make their clients happy, increase their commitment to the organization, and thus, resistance to change will diminish considerably and the results will be better. In this sense, it is necessary that we identify which factors make change easier.

2.4 Employer Branding

The concept of employer branding regards managing the company's brand by associating it with that of a "good" employer, by their employees at a given time, by the individuals that have worked in the company and by all persons that will be part of the organization in the future (Mascarenhas 2017).

In brand management, communication is essential and their positions towards different publics should be unique even though the attributes to be communicated might vary depending on who they are being communicated to. When it comes to employer branding, firstly it is necessary to define the employer value proposition (Mascarenhas 2017).

This should integrate the complete life cycle of each employee within the company from attraction, maintenance and separation (referring back to the definition of recruitment policies and practices, selection, remunerations, training and career development, etc.) (Mascarenhas 2017) and should be very attractive to the present and future employees.

Talented workers seek new and challenging professional experiences that are meaningful and impactful. Companies that bet on strong branding make it so that the brand (Inbranding) comes first in their workers' hearts and minds. Marketing and human resource departments work together to make it so that the workers live out the brand each day.

Companies intend their images to be positive through the satisfaction and well-being of their workers/employees. These organizations then become very competitive as places where everybody wants to work (Armstrong 2006, cit in Sokro 2012).

Within these circumstances, companies with a great reputation can easily attract, contract and keep the most qualified individuals which in turn will bring better results for the organization—this is the biggest outcome that employer branding promotes. A company with employer branding creates a kind of trademark in what comes to promoting its role as an employing entity.

3 Methodological Procedure

The empirical study focuses on a private and a public organization.

Documents from these organizations were analysed and semi-structured interviews that took place to individuals who are formally or informally responsible to put in place HRM practices.

The questions made sought to characterize organizations and their already existing practices, the practices that are being implemented currently as well as the assumptions that lead to the implementation of those practices.

The data analysis was complemented with information from similar studies to be able to establish comparisons.

4 Practices of Analysed Organizations

4.1 Private Organization

Twenty-year-old multinational technological consulting company was born in France with over six thousand tenured employees. In Portugal, it has about six hundred employees. Working in a very competitive field, this company has been promoting innovating human resource management practices which make it stand out and be more attractive. The organizational strategy is founded on the following structuring elements:

(i) Employee centric;
(ii) Focus on employer branding through constant and fun communication;

(iii) Organizational culture that supports innovation, creativity, sharing and involvement;
(iv) Happiness as a culture rather than a trend; the creating on the position of Happiness Manager is only one of the actions pertaining to that goal;
(v) Real focus on collaborator's well-being and happiness;
(vi) Importance of people and full confidence that happier people are more productive;
(vii) Living out the company as an experience and a bilateral relationship that must be constantly cultivated. From these, in what comes to human resource management practices, the following stand out.

4.1.1 Welcoming and Integration

The company starts off with a welcoming session. Every collaborator on Friday before joining the team is invited by the Happiness Manager to have a three hours long welcoming session.

The session is made up of breakfast with Happiness Manager, followed by a conversation with the CEO that welcomes them, the CEO that tells them about the company's history in Portugal and in the world, the Business Director that tells them about our current projects and clients in Portugal and the Director of Strategy and Human Resources who tells them about talent management at company.

After the session, the new employees take part in Friday's lunch, at the Lisbon headquarters with, on average, one hundred and fifty employees. On that day, we welcome them and the lunch allows to immediately begin the integration process since they can meet a large number of colleagues that will be part of their life with the company.

The Buddy Programme is already in the implementation process. Every new collaborator will have a buddy that will accompany them through their first year with the company.

The buddy will receive the collaborator on their first day of work and support them with any needed informal integration, showing them around and answering the questions we all want to know in the first workday—"where is the kitchen?".

4.1.2 Professional Training (Academia)

Framing: knowledge management focused on individual development allowing organizational development. Work on collaborator engagement through active participation in company life, assuming different roles like instructor, mentor, etc.

Internal core—encourages collaborator engagement, involvement with the company and in spreading knowledge.
Recognition—being internal instructor weighs on performance evaluation.

Compensation—every hour of training given is paid 150% of the collaborator's hour usual value.

Technological core—small communities of employees that share the technology with which they work: sharing of knowledge, clearing doubts and problem-solving amongst employees; networking; identification of training needs, potential internal instructors and identification of mentors and mentees; brainstorming and business opportunity; and identification of potential needs in ongoing projects or in new projects.

Quarterly training plan—constant evaluation of market and of collaborator's needs demanded a more flexible and nimble model: in Individual Development Plan meetings needs can be raised that will be communicated to the Talent Development team; after needs are analysed, a training catalogue is designed for the coming three months with training every Wednesday and Saturday; and employees have responses to their needs in a fast and timely manner.

Tailored training—can come out of the Individual Development Plan or from an isolated training request for one or more individuals, usually associated with a specific project being worked on or client; the best technical solution is evaluated by the Talent Development team (internal training vs. external training).

Meet and Tech—thematic informal monthly workshops held at the company's headquarters; food and drink and a facilitator who is a specialist in a given technological theme.

4.1.3 Performance Evaluation (Feedback and Coffee)

Already designed and in implementation process: well-defined feedback cycle with the clear intention of making feedback a part of the culture; employee centric; seeks to eliminate bureaucracy from the process and focus feedback on an Individual Development Plan that is fed both by those who lead and those who are lead; always looking to relate what the company expects from the employees (Performance and Potential) to what the collaborator expects from the company (Recognition and Opportunities).

4.1.4 Compensation and Benefits (Happiness Management)

Birth kit when collaborator's children are born; gift to the collaborator's children for children's day; birthday gifts to employees; postcard with personalized message on the anniversary of the collaborator's employment with the company; support to buy school material for each child at the beginning of each school year; free lunch every Friday; monthly event (every first Thursday of the month) with the goal of bringing together every collaborator for an informal Meet Up (such as Stand up Comedy; Karaoke; Fado night, etc.); big annual events (Popular Saints Festivities, Summer Sunset, September Reentré with collaborator's families, Christmas Party); company beneficiaries club in the intranet informatics platform where employees can access discounts on over 200 brands and have access to several exclusive benefit protocols for being from this company; when absent due to illness the Happiness Manager will send a personalized message and gift wishing the collaborator to feel better; referral prizes—the company pays 500€ for each referral of a contracted applicant (an incentive that generates engagement and commitment); monthly prize draw (with prizes such as summer festival tickets, movie tickets, etc.); and company satisfaction—quarterly satisfaction questionnaires where the opportunities for improvement are identified and any situation that may cause low satisfaction levels and may cause employees to leave is immediately acted upon.

4.1.5 Employer Branding

It involves communicating everything the organization does; inside (employees) to outside (social networks, etc.) looking to retain and attract; informal and fun communication style—using humour as a catalyser of good energy and positive image; making this company a reference in the way, it relates to every stakeholder with the main focus as the people who make up the organization; and total focus that organizational happiness is cultural and not a trend.

4.2 Case Study with a Higher Education Institution

In Portuguese Public Administration (PA), not unlike in most EU countries, there is a strong movement towards innovation and modernization (Lopes and Rodrigues 2010) which results in a new paradigm oriented towards public and consumer's interest. It has been several years since the public organization government system, based on the traditional administrative model, has stopped guaranteeing organization's efficacy (Bilhim 2017) and does not safeguard the conditions and unbiased needs of a public service.

Human Resource Management should, according to Weberian Theory (Pollit and Bouckaert 2004), count on: an impersonal recruitment and selection; a performance evaluation—understood as a mutual feedback and opportunity for transparency; a continuous professional training plan capable of making each professional a specialist in their field and a remuneration based on merit shown by performance evaluation as well as the organization's overall performance.

New participative models of valuing people are emerging in PA (Mozzicafreddo and Gouveia 2011), especially in leadership, communication, motivation and satisfaction, bringing about a sense of pride and belonging, positive perception of work and more dedication and engagement. These new models that stay away from the usual bureaucracy of PA are designed in a way that tries to adapt to the reality of each institution, considered good practices, they are promoting autonomy, initiative and reducing the levels of subordination with a focus on valuing of competencies and increasing on the added value of work.

This case study with a higher education institution shows a changing culture in what comes to the management of people and a human resource management with a new mission of promoting human resource practices that are socially responsible, shared and sustained by those in charge, generating development, workers' qualification and motivation and a way of generating value for the institution.

At the end of 2017, this higher education institution employed 703 people with the following distribution: 533 teaching staff and 156 non-teaching staff. It also had seven research scholars. Regarding education levels, 32% held a doctorate degree, 27% masters, 27% an undergraduate degree and 14% 12th grade or under. The average age is 45 with 56% of them being female and 43% male.

To put into practice this new approach to People Management, a programme called "Desenvolver+" [Develop+] was created with the goal of making activities that promote strategic development available—developing, qualifying and motivating people through four structuring branches: (i) competency development; (ii) recognition and merit; (iii) workers' well-being; and (iv) workers' participation.

Branch 1: competency development, with training programmes with a strong behavioural component, developing soft skills and transformational leadership and collaboration competencies.

The plan is annual but has flexible characteristics, to be able to respond to workers' and team challenges, with new actions being put in place according to the needs that emerge. The tendency in this institution is to lead workers to develop their competencies by giving them more autonomy in how they spend their time, ultimately making them responsible for their own development.

Training is by instructors both from within and from outside the organization. Teaching staff participate actively as internal instructors, and this contribution is shown in their performance evaluation. In what comes to training, there are monthly thematic workshops open to the entire community called "Ciência à Conversa" [Talking Science] which happens at lunchtime and is led by specialists focusing on themes of collective interest.

Branch 2: recognition and merit, with activities that aim to recognize professionals with excellent performance, creating of good practice projects—recognition of teams with projects relevant for the institution; workshop developments about "compliment and a positive attitude at work" with a goal of creating a positive feedback culture.

Comprises activities that recognize professionals that have shown exception behaviour, zeal and competency in performing their functions (Excellency medal) that have been performing their functions for over 20, 30 and 40 years of service (dedication medals), or for workers who are retiring, as a way of distinguishing the services they provided through their active years (recognition medal). It also comprises the formal recognition of teachers who have concluded their doctorate degree or who have obtained a specialist title (qualification medals);

Branch 3: workers' well-being, with activities connected to health (for example nutritionist appointment, acupuncture, herbalism, healthy food and cooking workshops, worker's physical condition evaluation and action plan development, with partnerships with local food co-ops to get more sustainable food); free sports and the institution's gym, physical activities (for example, hikes, city by bicycle); improvement of physical conditions of work, new equipment, better safety conditions, cultural activities (theatre, historical knowledge from the region, dolphin and bird watching); stress reduction activities (mindfulness, time management workshops); organization climate studies and monitoring of results in different sectors—leadership, colleagues, work condition, internal communication and social support.

Branch 4: workers' participation, integrating idea and complement presentations, internal participation models implementation that pertain to working procedures and management mechanisms, with participatory meetings and worker representation in different organs of the organization.

4.2.1 Results Obtained

The project "*Desenvolver+*" is changing human management. By focusing on human resource management at its best, especially looking to develop skills, motivation and team participation, the project promoted closeness and a sense of partnership amongst the different services and schools, which came directly with an improvement of the services provided to the academic community.

It is worth highlighting: (i) an increase in average training hours (27 h/year) as well as the number of workers involved (94%/year); (ii) an improvement in the functioning and articulating of services; (iii) improvement of internal commitment; (iv) increasing cooperation among services and projects number across departments; (v) increase in service satisfaction level; (vi) and increase in number of students—from 5053 in 2014 to 5726 in 2018.

As examples of the results achieved, some testimonies from the key players of good practice are given:

> IPS, with their program *Desenvolver+* has been shifting their paradigm of policies and practices of HR, promoting activities that have been successful with the recognition of non-teaching staff as well as their competency development. (employee A—highly qualified employee)
>
> Professional training should be adapted to the specific functions and strategies the organizations hold, aiming to guarantee and updating and acquiring of knowledge; the *Desenvolver+* program, besides ensuring that, has been promoting team building and team spirit activities. These activities resulted in a strengthening of bonds and synergies amongst IPS workers, which has been translated in a healthy cooperative environment which has been very gratifying and fundamental for my development as a professional and as an individual. (employee B—employee union delegate)
>
> *Desenvolver+* encourages HR development at IPS. Everybody knows an participates in the plan and in the mission of IPS, being challenged and encouraged to achieve new strategic objectives. It highlights the importance of professional training, the valuing of people, their motivation and well-being. (employee C—Supervisor)

4.2.2 Critical Success Factors

The success of this programme comes out of a strong adherence from the higher management to this new way of managing people, especially through availability to listen, change and transform. This success also results from a capacity to identify, anticipate and respond to the needs of employees—both through questionnaires and through in person meetings with higher management and workers, finding solutions that align with the new HRM strategy.

Other positive factors that are worth noting are: a clear strategic direction and following of the multidimensional programme, fulfilling expectations that were there at its inception, the step-by-step sharing of results with the workers; the training and preparing of leaders in what comes to management of change of people and the creation of a compliment and positive feedback culture; the involvement of everyone in training programs and team building events, the reinforcement of bonds and construction on a corporate identity; and the close accompaniment of teams.

5 Conclusion

In the text, we presented we sought to show that human resource management practices can be determinant for the promotion of organizational creativity and innovation. Regarding core human resource management, the essential aspects and objectives have remained the same. What changes is the path chosen for those objectives to be achieved, with the employees apparently taking on centre stage, under the assumption that if employees are satisfied and happy, they will be more pleased and enthusiastic about fulfilling their functions and producing increasingly positive results for organizations.

What we verified both in the private company and in the public organization were changes that might even seem like details but in what comes to the people making those changes, they can make a difference and contribute to the well-being of employees thus to their commitment and effort, leading to higher organizational competitiveness and performance.

We would encourage and hope independent studies are complemented to collect information with these organizations' employees to confirm what is mentioned by those in their leadership.

References

Alis, D., Des Horts, C. H. B., Chevalier, F., Fabi, B., & Peretti, J. M. (2012). *Gestão dos Recursos Humanos, uma abordagem internacional*. Lisboa: Edições Piaget.

Amabile, T., Conti, R., Coon, H., Lazenby, J., & Herron, M. (October 1996). Assessing the work environment for creativity. *The Academy of Management Journal, 39*(5), 1154–1184.

Amaral, J. M. R. (2015). Criativos, precisam-se? Criatividade e Práticas de GRH. *Dissertação de Mestrado*. Lisboa: ISEG.

Armenakis, A. A., & Harris, S. G. (2009). Reflections: Our journey in organizational change research and practice. *Journal of Change Management, 9*(2), 127–142.

Baranano, A. M. (2005). Gestão da Inovação Tecnológica: Estudo de Cinco PMEs Portuguesas. *Revista Brasileira de Inovação, 4*(1), 57–95.
Barbieri, J. C. (2012). Organizações Inovadoras Sustentáveis. *Caderno de Inovação, 3*, 5–9. Fundação Getúlio Vargas.
Barney, J. B., & Wright, P. M. (1998). On becoming a strategic partner: The role of human resources in gaining competitive advantage. *Human Resource Management, 37*(1), 31–46.
Bilhim, J. (2017). Nova governação pública e meritocracia. *Sociologia, Problemas e Práticas* (84), 9–25.
Combs, J., Hall, S., Kitchen, D., Liu, Y. (2006). How much do high-performance work practices matter? A meta-analysis of their effects on organizational performance. *Personnel Psychology, 59*, 501–528.
Curado, C. (2018). Human resource management contribution to innovation in small and medium-sized enterprises: A mixed methods approach. *Creativity and Innovation Management, 27*, 79–90.
Dietrich, A., & Pigeyre, F. (2016). *La Gestion des ressources humaines* (3rd ed.). Paris: La Découverte.
Lawler, E. E., Boudreau, J. W., & Jamrog, J. (2010). *Impact of recession: What has happened to HR?* (pp. 1–8). Center for Effective Organizations Publications.
Lopes, A., & Rodrigues, C. (2010). A modernização da administração pública passa por uma revolução burocrática. Lusíada. *Economia & Empresa* (10), 95–139. Lisboa.
Lukkap, P. (2014). Presente e Futuro do mercado Laboral. In V. Meirinhos & A. C. Rodrigues (orgs.), *Gestão e Desenvolvimento de Recursos Humanos, Tendências e Boas Práticas* (pp. 11–30). Vida Económica: Porto.
Mabey, C., & Salaman, G. (1995). *Strategic human resource management*. Oxford: Blackwell Publishers.
Mascarenhas, B. G. (2017). Employer branding no ciclo de relacionamento com empregados. *Comunicação com líderes e empregados, 3*, 103–108.
Martins, L. F. (2016). A Influência das práticas de Gestão de Recursos Humanos na criação de Inovação: Estudo de caso. *Dissertação de Mestrado*. Braga: Universidade do Minho.
Melo, P., & Machado, C. (2015). *Gestão de Recursos Humanos nas Pequenas e Médias Empresas, Contextos, Métodos e Aplicações*. Lisboa: RH editora.
Mozzicafreddo, J., & Gouveia, C. (2011). Contextos e etapas da reforma na administração pública. In J. Mozzicafreddo & S. J. Gomes (orgs.), *Projetos de Inovação na Gestão Pública* (pp. 5–62). Lisboa: Mundos Sociais.
Nohria, N., Groysberg, B., & Lee, L. E. (July 2008). Employee motivation: A powerful new model. In *Harvard business review* (pp. 1–8).
OECD. (2005). *Oslo manual, guidelines for collecting and interpreting innovation data* (3rd ed.). OECD publishing, European Comission.
Pedro, M. D. L. C. (2015). *A GRH em contexto de crise: A centralidade da perceção do declínio organizacional* (Tese de doutoramento). ISCTE, Lisboa.
Pollit, C., & Bouckaert, G. (2004). *Public management reform. A comparative analysis*. Oxford: Oxford University Press.
Rodrigues, A. F., & Veloso, A. (2013). Contributos da Gestão de Recursos Humanos para a Criatividade e Inovação Organizacional. *Revista de Psicologia Organizacional e do Trabalho, 13*(3), 293–308.
Sokro, E. (2012). Impact of employer branding on employee attraction and retention. *European Journal of Business and Management, 4*(18), 164–173.
Sousa, F. C., & Monteiro, I. P. (2017). *Liderança de equipas na resolução de problemas complexos, um guia para a inovação organizacional* (2ª edição). Lisboa: Sílabo.
Ulrich, D., & Brockbank, W. (2005). *The HR value proposition*. Boston: Harvard Business School Press.
Wright, P. M., & McMahan, G. C. (1992). Theoretical perspectives for strategic human resource management. *Journal of Management, 18*(2), 295–320.

Employer Branding: Issues of Tailoring Your Message in the Modern Age

Nick G. Chandler and Tamas Nemeth

Abstract Although differentiating which employers are the best to work for is not a new concept, employer branding has been in greater focus since the term was first used in 1996. More recently, increasingly sophisticated communication methods, technology and services have all changed the environment in which the employer brand is devised, developed and communicated. In this chapter, we first define employer branding and present its theoretical framework. We then examine the key aspects of employer branding that have been adapted or will adapt to the modern age.

Keywords Employer branding · Employer value proposition · Offering · Communication

1 Defining Employer Branding

Ambler and Barrow (1996) first coined the term and defined the concept of employer branding as: "The package of functional, economic, and psychological benefits provided by employment and identified with the employing company" (p. 187). Since 1996, a number of companies have focussed on building the employer brand, such as Shall and Unilever. Backhaus and Tikoo (2004) emphasise that the employer brand is concerned with differentiation from competitors.

A distinction must be made between the employer brand and the corporate brand. Barrow (2007) points out that the employer brand involves the company's identity as an employer for both employees and potential applicants from the labour market, i.e. has both internal and external focus. This is also highlighted by recent studies such as that of McLeod and Waldman (2011) who suggest employer branding involves external and internal actors perceiving the organisation as a "great place to work". Corporate brands, on the other hand, have a generally external focus.

N. G. Chandler (✉) · T. Nemeth
Budapest Business School, Institute of Management, Budapest, Hungary
e-mail: Chandler.nicholas@uni-bge.hu

C. Machado and J. P. Davim (eds.), *Entrepreneurship and Organizational Innovation*, Management and Industrial Engineering, https://doi.org/10.1007/978-3-030-19289-1_3

The importance of employer branding cannot be overstated. The perception of the workplace as a great place to work has been found to have an impact on employee retention (Cole and Burch 2006).

2 Conceptual Framework

An experience involves two key elements: the employer value proposition (EVP) and the employee experience. The value proposition of a company is developed using information such as the working environment (e.g. leadership, organisational culture) and the reputation of the company as a place to work (Sullivan 2002). The employee experience in turn affects the reputation of the organisation and covers the employees' entire working life. Michaels et al. (2001, p. 43) defined that EVP is "the holistic sum of everything people experience and receive while they are part of a company—everything from the intrinsic satisfaction of the work to the environment, leadership, colleagues, compensation and more".

Hubschmid (2012) considers the EVP approach as first defining the target audience and then segmenting this audience into groups. For example, undergraduates will have preconceptions of a "great place to work" but may differ greatly from potential employees with substantial experience, in terms of values, expectations and perspectives. This is also the case for grouping existing employees based upon occupation, or level in the hierarchy, for example.

A recent approach to employer branding considers the entire employment life cycle and also considers all stakeholders (internal and external). This is referred to as the "community concept" and has been adopted by companies such as Mars, P&G and Google. This approach requires an audit of all stakeholders as to what they perceive as the company brand, on the one hand, and an examination of employee experience, on the other (Rosethorn 2009). By combining these two audits, leaders can understand how the brand is perceived by all stakeholders and what areas can be amended or developed to change the employer brand.

Following a review of stakeholder perceptions, values and expectations of the organisation as an employer, according to their groupings, consideration must be given to the current offering of the organisation and the capacity of the organisation to meet the current demands of the stakeholder groups, i.e. the changes that should be made to the current offering in order to attract the appropriate group(s). Finally, the form of communication needs to be considered based upon current practices, potential employee preferences and organisational capabilities. Each of these considerations will be dealt with in the following sections of this chapter, with special attention to the most recent approaches and innovations for each element of effective employer branding.

3 Perceptions of the Organisation

The perceptions by stakeholders of the organisation as a workplace are crucial in understanding the starting point from which the employer brand can be built or developed. Parment and Dyhre (2009) suggest some of the most common attractive employer characteristics as:

- quality working relationships (trust, respect, self-worth and recognition)
- leadership (e.g. preferred style, followership)
- participation (decision-making, autonomy)
- clear values (awareness and understanding of core values of the organisation, as well as regular feedback)
- meaning and fun (a mission and vision at all levels, coupled with enjoyment of the job and the workplace).

Cafolla (2008) also examined attributes that attract an employee to apply for a position and found that although salary was once the main incentive, things have since changed—elements such as the opportunity to work for a well-branded company, developmental aspects such as training (especially overseas, career growth), as well as an attractive work environment and social status, had superseded salary as the key motivator to apply. In a study by Terjesen et al. (2007), it was found that the five most important organisational attributes were: "invest heavily in the training and development of their employees", "care about their employees as individuals", "clear opportunities for long-term career progression", "variety in daily work" and "dynamic, forward-looking approach to their business".

Although the attributes seem to be clearly set, they are likely to vary from one target group to another—attractiveness varies according to age, gender, educational background and cultural characteristics (Hubschmid 2012). Terjesen et al. (2007) found that differences existed by gender, with female participants having a tendency to prefer attributes the following attributes in comparison with their male counterparts: "really care about their employees as individuals", "variety in your daily work", "friendly, informal culture", "employ people with whom you feel you will have things in common", "use your degree skills", "relatively stress-free working environment", "internationally diverse mix of colleagues" and "require you to work standard working hours only". Conversely, men rate just one attribute as more important: "a very high starting salary".

Furthermore, attractiveness itself can be considered an attribute if it is generally held in society or a certain group that a company has a good reputation as a place to work, as it has been found that this increases the attractiveness for other individuals as well (Edwards 2010), creating a form of "snowball effect".

Finally, another attribute that can attract a potential employee is due to the increased stress on companies to concern themselves with sustainability and socially responsible practices, as companies heavily involved in such practices have been found to be more attractive (Turban and Greening 1996).

4 The Employee Experience

The employee experience helps create values and influence (Edwards 2010) of both external and internal stakeholders. The experience provided and valued by employees is communicated to existing and potential employees through a number of forms such as through company-promoted advertising of the workplace, through independent sites reviewing employers (e.g. Glassdoor) or through word of mouth in professional networks or the local community (McLeod and Waldman 2011).

A number of models have been suggested for managing the employee experience in relation to employer branding, such as Barrow and Mosley's (2005) twelve-stage model or McLeod and Waldman's (2011) four-stage model. We will consider the later in this chapter, which developed steps in the employee lifecycle as a means of adopting the community approach mentioned earlier in this chapter: attract, integrate, perform, retain and transition, which are as follows:

1. Attract: Ensuring the best candidates want to work for a company. This involves the auditing of the potential and current employees to assess brand awareness, and current company reputation. During the interview process, the company brand can be communicated (including those attributes classed as key to attracting the suitable target group) and it is also an opportunity to assess the perception of the employer by the job candidates.
2. Integrate: The new employee is integrated into the team. The job tasks are allocated, and over time the new employee learns the norms and values of the company culture. As the employee learns and integrates, they contribute to the employer brand.
3. Perform: The employee has been working in the job for some time and has certain expectations that were communicated based on previous perceptions prior to joining the company, and those acquired during the attraction and integration stages, such as learning and development. The meeting of these expectations will affect the engagement of the employee and satisfaction, which in turn will affect how the employee projects the image of the organisation to internal and external stakeholders.
4. Transition: Through internal promotions or redeployment a person "transitions" to a new position or, for whatever reason, may decide to quit the company. In the former case, the cycle restarts at the integration phase, and in the latter case, the person has been attracted to another company and so the process begins again at the first stage.

Earlier in this chapter, we found that a key part of the early stages of employer branding involves the segmentation of external stakeholders. Based on an empirical study, we suggest the segmentation of employees as well—especially if we consider the entire employee experience over time.

If we take a longitudinal view of employee experience, then we must consider three central aspects: first, values, expectation and perceptions change over time; second, in larger organisations a natural segmentation occurs on the basis of values

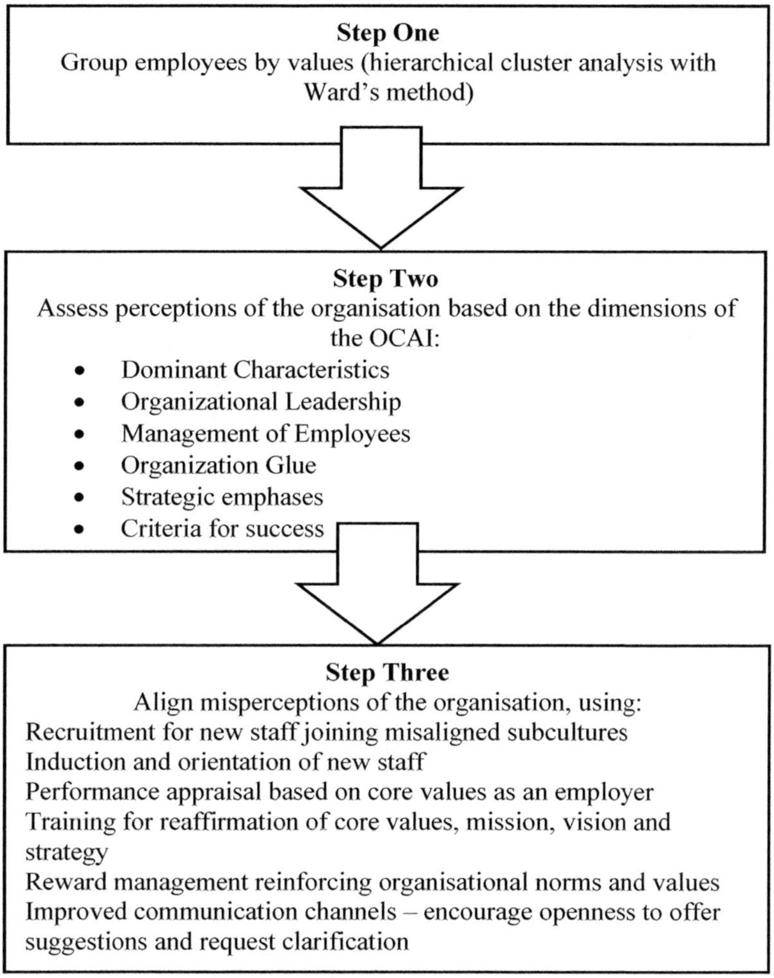

Fig. 1 A model for segmentation of employee perceptions of the employer brand

and expectations in the form of subcultures; and third, these subcultures are prone to distinct differences in values and perceptions (Martin 2002), sometimes resulting in misperceptions of the organisation (Chandler et al. 2018). Thus, we suggest the following model using the Organizational Culture Assessment Instrument (OCAI) developed by Cameron and Quinn (1999), as this covers values which can be used for clustering subcultures and perceptions, which can be used for assessing the perceptions by employees of the organisation Fig. 1:

Using this model, allowances are made for differences in employee experiences and perceptions of the employer, and misperceptions can be understood and targeted for alignment (if necessary) based on particular subcultural groupings. This model also allows management to see in greater detail the areas where the employer brand needs attention based on employee experience (see dimensions in step two).

Step three of this model also considers how employee engagement figures in the concept of employer branding. The reference to communication of mission, vision and strategy is based on the finding by Rosethorn (2009) that these elements affect the employees' willingness to stay. Clear communication of the mission and values will lead to an emotional attachment to the organisation and thus greater employee engagement. An understanding and acceptance of the vision will lead to a vested interest and concern for the future of the company and lead to greater engagement as the employee commits to the future success of the organisation. The tools for aligning perceptions (and thereby increasing engagement) are based on the four components of the engagement model of Rosethorn (2009): corporate reputation and brands; rewards and recognition; culture and environment; and opportunity (to grow and develop). When considering the content for the tools used in step three, Bersin (2015) suggests some useful areas for adopting as areas for improving the employee brand: meaningful work, supportive management, positive work environment, growth opportunity and trust in leadership. These tools for improving perception with existing employees and choice of attributes to attract potential employees can only be used if the organisation has the capacity to do so, and this will be covered in the following section of this chapter.

5 The Offering of the Organisation

The organisational can only control certain elements of how it is perceived by job seekers and employers in terms of its weaknesses and strengths. Some elements are freely available to the public, at least in the case of large organisations, such as profitability. Edwards (2010) examined how profitability is a key factor that helps predict job seeker's perception of an organisation.

The offering itself is also seen as an indicator of the strength of the company in general and the combination of salary, benefits and rewards as indicators in particular. Fractl conducted a survey in 2017 of 2000 workers in the USA and found that salary itself came second to better benefits, such as health, dental and vision insurance, for 88% of the sample.[1]

A similar number of respondents indicated they would switch jobs if working hours were more flexible and this is distinct indicator of the need for companies to move with the times in their offerings to both job seekers and current employees. In the world of work, there is an increasing desire for "quid pro quo"; i.e. if people are expected to have meeting with their overseas counterparts before seven in the morning or late into the evening, then the same degree of flexibility expected of the employee should also be expected of the employer. Thus, the need for flexibility is

[1] https://hbr.org/2017/02/the-most-desirable-employee-benefits.

also centred around the need for mutual respect and trust, both of which can also be considered a part of the employer brand.

The offering of the organisation also needs to move with the times in terms of technology. In the modern age, we are used to most aspects of daily life enabling us through technology to have the information we need when we want and where we want, as well as deeply personalised. Some companies are also ensuring these aspects are included in the company offerings, such as Lyra Health. One of this company's employee benefits is behavioural health case, and they ensure that employees can easily access practitioners that cater to their specific needs, rather than simply having the benefit as part of the employee package. The role of technology in employer offering can also be seen in employee benefits such as egg freezing and in vitro fertilisation.[2]

As a final point on the topic of technology, as big data and automation speed up communication flows, the method of communication is a key concern of employer branding. As artificial intelligence and VR play an increasing role in the recruitment process, such as in the British Army,[3] there may well be ways for virtual reality too to be a means for communicating the employer brand to employees and job seekers alike. The communication of the employer brand will be covered in the following section.

6 Communication of the Brand

Effective communication is a vital prerequisite to success for high-performance organisations. Yet, it is a commonplace that communication is, in most cases, weak. Employees complain that they do not know and do not understand corporate priorities while frustrated executives firmly assert that they have spent a lot of time investing in and explaining their priorities. There is countless evidence that upward communication from the lower levels is defective as subordinates (but even lower-level leaders) are afraid to express their true opinion to the higher levels of hierarchy. Consequently, senior management is often unaware of the organisation's communication dynamics. And it is also true for job seekers versus recruiters, since they both look for competitive advantages, best practice, creative attitude and, surely, interest-centred elements. These needs and features of companies and individuals immerse in the very field of employer branding.

The place of work has changed quickly over the years and will accelerate even more. According to Upwork's 2018 Future Workforce Report, nearly three times as many HR managers felt hiring had gotten harder in the past year than easier and 96% of them had open positions.[4] This broadened the way in front of previously existing methods of working (e.g. remote access, global teams, flexible hours, paperless

[2]https://www.emotivebrand.com/employer-brand/.

[3]https://visualise.com/case-study/british-army-vr-recruitment-experience.

[4]www.upwork.com.

offices)—all of this with an increasing number of communication channels both in personal and corporate life.

Thus, apart from work-related, task-related, value-related employer branding elements, communicating on expectations, behaviours and necessities has come to the light: there are no one-size-fits-all solutions any longer. The different generations of workforce and the different stages of corporate development and communication capability maturity resulted in a really diversified labour market phenomenon (Ilyas et al. 2018).

7 The Twofold Nature of Communicating Employer Branding

Backhaus and Tikoo (2004, p. 502) argue that employer branding includes promotional activities of the company as an employer both externally and internally (the purpose is to attract new recruits and make sure their current employees become engaged with the company's culture). Since employer branding is based on a company's culture and values, companies should clearly communicate what they can offer to future employees, as well as existing employees and customers (Foster et al. 2010), as shown in Fig. 2:

Research reveals that the areas of employer branding are not mutually exclusive, but instead an intertwisted mixture of branding issues that form the corporation's core values (Näppä et al. 2014). Thus, it seems evident to have cooperating divisions for jointed efforts: HR and marketing. In order to fulfil tasks and reach the desired outcome, these functional units should agree on a communication strategy and complement each other's work. For instance, if an external audience might be interested in an internal matter, marketing will communicate it; if an internal audience might find an external matter intriguing, HR will communicate it. As a consequence of the twofold nature of employer branding communication, when a customer has a particularly good expe-

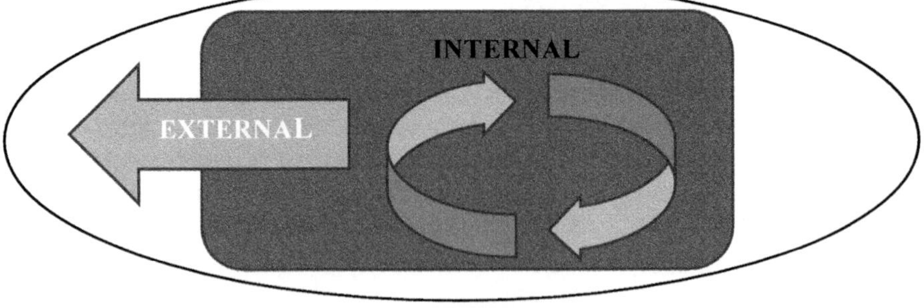

Fig. 2 Scope of employer branding communication

rience with the organisation and uses social media to communicate the appreciation, this may lead to greater awareness and more potential customers.

There is a consensus that the medium in which marketing and HR may collaborate on communicating a shared brand identity and customer experience is social media. Social media can connect the two experiences since it is a form of communication that can reach across the management–employee–customer divide (Whiting and Deshpande 2016) Social media act as a sort of personal information source and word of mouth. File et al. (1992) discuss the importance of word of mouth in creating a positive image among a company's customers and employees as well.

This, together with the recruitment activities, makes up a holistic corporate communication strategy. This can best be obtained as a first step by carrying out a communication audit in order to see an objective picture of the company (what is happening, through what channels, what intensity, what frequency, whose involvement, etc.). The results of a communication audit provide reliable feedback, so leaders can make proper decisions about where to change existing (internal and external) practices.

A new trend has taken shape in recent years, i.e. to create formal public relations messages in a more effective and credible way by enabling employees to act as brand advocates or informal spokespersons. They might use their personal links and networks to spread positive word of mouth to promote and defend the organisation's brand in public.

For a decade now, employers have had to deal with issues concerning employees' Internet use via organisational information and communication technology (ICT) infrastructures. Although lots of organisations block these sources of information (i.e. Google, Facebook, Instagram, YouTube, etc.), others let them into the very heart of internal communications channels and create platforms for the so-called social Intranet. The great dilemma around this decision lies in whether to associate danger with the nature of content and the hazards of being online (viral attacks for instance), but on the other hand, the option to discard this type of communication and collaboration way would surely deterrent for Gens Y and Z employees.

8 Towards External Sources

Since the introduction of social media, the corporate communication landscape has changed significantly, and thus organisations need to find new innovative ways to communicate: the former one-way communication between organisations and stakeholders has changed into a multidimensional two-way communication (Miles and Mangold 2014; Schivinski and Dabrowski 2016). The corporate communication landscape has thus changed from a single controlled source to countless touchpoints with the outside world. This is nothing less than a dramatic shift in paradigm (Kohli et al. 2015). Social media has become a much criticised yet substantial part of employees' everyday lives, and organisations need to understand that their activity on social media platforms has the power to influence the corporate reputation. What's more, communication departments cease to exist as only sources that manage corporate

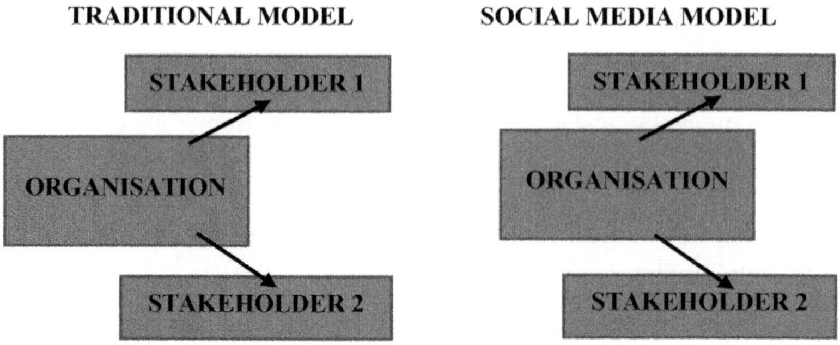

Fig. 3 Traditional and the social media model. *Source* Kohli et al. (2015)

reputation, and any new communication strategy should be developed accordingly (i.e. managing risks and utilising benefits of social media as seen in the examples with Dell, Unilever and GE (Dreher 2014; Miles and Mangold 2014) (Fig. 3).

Five of the most used social media platforms globally are **Facebook, LinkedIn, Twitter**, **Instagram and blogs** (Statista 2018). While users on Facebook display relational information, interest and hobbies on their profiles, LinkedIn users display professional information in the format of a CV, so this is more of a professional networking site where recruiters and professionals looking for job opportunities may meet. Another popular social media platform is Instagram where users share photographs and videos through their mobile phones.

Twitter, often described as a microblog, also allows users to write posts, but within a 140-character limit—providing a quick posting platform to reach many interested people. A blog is defined as an online diary where users, i.e. bloggers, frequently publish posts on a website or online page that allows comments from readers, which enables a two-way communication. Blogs can be seen as reliable channels of communication, and lots of organisations construct platforms for blogs in order to reach stakeholders. It is more than highly recommended that communication departments should provide content for employees on these platforms.

Brand advocacy is often described as a kind of **word-of-mouth communication** where brand advocates endorse an organisation by voluntarily giving recommendations to other stakeholders (Fullerton 2011). Brand advocacy is not a new concept, but the term for brand advocates differs throughout literature; the most common are brand ambassadors, brand champions, brand evangelists and brand maniacs (Morhart et al. 2009).

Gamification, i.e. the use of game design elements in a non-game context in order to create a competitive environment, is an increasingly popular method to obtain job seeker interest, employee engagement and enhance advocacy. Gamification might also serve as a tool for encouraging employees to use social media (Dreher 2014).

9 Towards Internal Sources

When it comes to employer branding, any management should use a two-way communication pattern: turning to the outside world (the labour market) and, at the same time, to the internal one as well. This latter one can lead to enhancing brand-specific skills and brand performance of employees (Punjaisri et al. 2009). Miles and Mangold (2005) divide internal employer branding communication into the following:

- formal internal communication (such as memos sent by HR department)
- informal internal communication (such as interaction with co-workers)
- formal external communication (such as advertising)
- informal external communication (such as customer or supplier feedback).

The different types of internal communication indicate that employees gain information not only from within the organisation but also through external channels of communication, which, in turn, influence the overall organisational identity (Hatch and Schultz 2002). We argue that internal brand communication should not be limited to the marketing department, although this is a common understanding in the literature (Punjaisri et al. 2009); it must be spearheaded by the HR department and also be initiated by employees, especially brand champions (Thomson et al. 1999).

The main goal is to communicate the principles and values of organisations in order to build and then strengthen employees' sense of belonging (De Roeck et al. 2014; Asha and Jyothi 2013). This need to involve the well-planned and efficiently implemented HR aces, i.e. wages and benefits, employee empowerment, quality of working life, organisational socialisation, employee satisfaction, and training and talent development activities, among others. Working on and developing these key factors can be executed in a series of internal communications platform, channels and methods, as shown in Fig. 4:

It is highly significant to drive employer branding internally by creating channels for the "voice of the employee"—to allow employees to provide feedback. Ways of doing this can certainly be done through a two-way communication between management and employees, employees and employees. This is a challenge that requires careful planning (definition of timings, goals, audiences, budget), researching (internal climate research and cultural fit analysis), integrating (aligning internal and external messages, business goals with brand values, defining the visual message and communication tone), facilitating (decide on degree of staff empowerment, obtain and sustain staff involvement), implementing (using the various existing communication channels to communicate at all hierarchical levels), quantifying

ELECTRONIC	PERSONAL	ADDITIONAL
E-Mails	Meetings	Corporate Programmes
Newsletters	Town Halls	Hr Protocols
Intranet	Word-Of-Mouth	Leadership

Fig. 4 Channels and methods of internal communications

(employee engagement with the brand), reacting (providing regular feedback) and changing (continuously improving) (Mahnert and Torres 2007: 58). This latter group of activities should be seen as a powerful HR toolkit that enables HR (and indeed the management) to communicate with the whole organisation and job seekers in an integrated and value-rich way, in short, effectively.

10 Chapter Review

In this chapter, we have examined the key elements of employer branding and how they are being adapted in the modern age. We have seen the increased transparency of the employer environment through sites such as glassdoor.com and the drive for greater accessibility and availability of benefits due to technological advances. Moreover, the message developed and communicated by employers quickly meets with perceptions of job seekers' and employees' realities.

Many of the drives for change in the modern age appear to be a catch-up with aspects already familiar in everyday life and so can be considered as enforced evolution rather than revolutionary steps. However, with examples such as the use of virtual reality by the British Army in the recruitment process, it seems that the employers are bridging the gap quickly and strive to be different themselves from one another as "great places to work".

We have found that employer branding has many sides and that the inherent complexity is not reduced, but rather increased by changes in the modern age. The speed of information transfer and transparency have all led to a changing view of the employer brand: employee issues cannot be hidden, and perceptions can be quickly affected, leading to a need for a collaborative and integrative approach by HR practitioners.

References

Ambler, T., & Barrow, S. (1996). The employer brand. *Journal of Brand Management, 4,* 1–22.

Asha, S., & Jyothi, P. (2013). Internal branding: A determining element of organizational citizenship behaviour. *Contemporary Management Research, 7*(1), 37–57.

Backhaus, K., & Tikoo, S. (2004). Conceptualizing and researching employer branding. *Career Development International, 9*(5), 501–517.

Barrow, S. (2007). *The future of employer branding and HR? Employer branding: The latest fad or the future for HR?*. London: Chartered Institute of Personnel and Development (CIPD) Guide, CIPD.

Barrow, S., & Mosley, R. (2005). *Bringing the best of brand management to people at work*. Wiley: The Atrium, Southern Gate, Chichester, England.

Bersin, J. *Predictions for 2015 research report: Redesigning the organisation for a rapidly changing world*. https://legacy.bersin.com/uploadedfiles/. Accessed November 12, 2018.

Cafolla, L. (2008). How to build an effective employer brand. *China Staff, 14*(9), 23–26.

Cameron, K. S., & Quinn, R. (1999). *Diagnosing and changing organizational culture based on the competing values framework*. MA: Addison-Wesley Publishing.
Chandler, N., Csepregi, A., & Heidrich, B. (2018). *The world I know: Knowledge sharing and subcultures in large complex organisations*. In E. M. Vătămănescu & F. Pînzaru (Eds.), *Knowledge management in the sharing economy edition: Knowledge management and organizational learning* (Vol. 6). Springer Editors. https://doi.org/10.1007/978-3-319-66890-1_7.
Cole, M., & Bruch, H. (2006). Organizational identity strength, identification and commitment and their relationship to turnover intention: Does organizational hierarchy matter? *Journal of Organizational Behavior, 27*, 585–605.
De Roeck, K., Marique, G., Stinglhamber, F., & Swaen, V. (2014). Understanding employees' responses to corporate social responsibility: Mediating roles of overall justice and organisational identification. *International Journal of Human Resource Management, 25*(1), 91–112.
Dreher, S. (2014). Social media and the world of work: A strategic approach to employees' participation in social media. *Corporate Communications: An International Journal, 19*(4), 344–356.
Edwards, M. R. (2010). An integrative review of employer branding and OB theory. *Personnel Review, 39*(1–2), 5–23.
File, K. M., Judd, B. B., & Prince, R. A. (1992). Interactive marketing: The influence of participation on positive word-of-mouth and referrals. *Journal of Services Marketing, 6*(4), 5–14.
Foster, C., Punjaisri, K., & Cheng, R. (2010). Exploring the relationship between corporate, internal and employer branding. *Journal of Product and Brand Management, 19*(6), 401–409.
Fullerton, G. (2011). Creating advocates: The roles of satisfaction, trust and commitment. *Journal of Retailing and Consumer Services, 18*(1), 92–100.
Hatch, M. J., & Schultz, M. (2002). The dynamics of organisational identity. *Human Relations, 55*(8), 989–1018.
Hubschmid, E. (2012). *Shaping efficient employer branding strategies to target generation Y*. Bern: Peter Lang AG.
Ilyas, M. A., Shaari, H., & Alshuaibi, A. S. I. (2018). Transformation of the employees into brand advocates through employer branding. *International Journal of Academic Research in Business and Social Sciences, 8*(6), 361–370.
Kohli, C., Suri, R., & Kapoor, A. (2015). Will social media kill branding? *Business Horizons, 58*(1), 35–44.
Mahnert, K., & Torres, A. (2007). The brand inside: The factors of failure and success in internal branding. *Irish Marketing Review, 19*, 54–63. Special Issue on Irish Perspectives on Marketing Relationships and Networks.
Martin, J. (2002). *Organizational culture: Mapping the terrain*. Thousand Oaks, CA: Sage.
McLeod, C., & Waldman, J. (2011). *The HR trailblazer: Unlock the potential of your employer brand in a digital age*. E-book: http://www.eBookIt.com. Accessed November 12, 2018.
Michaels, E., Handfield-Jones, H., & Axelrod, B. (2001). *The war for talent*. Boston, MA: Harvard Business School Press.
Miles, S. J., & Mangold, W. G. (2005). Positioning Southwest Airlines through employee branding. *Business Horizons, 48*, 535–545.
Miles, S. J., & Mangold, W. G. (2014). Employee voice: Untapped resource or social media time bomb? *Business Horizons, 57*(3), 401–411.
Morhart, F. M., Herzog, W., & Tomczak, T. (2009). Brand-specific leadership: Turning employees into brand champions. *Journal of Marketing, 73*(5), 122–142.
Näppä, A., Farshid, M., & Foster, T. (2014). Employer branding: Attracting and retaining talent in financial services. *Journal of Financial Services Marketing, 19*(2), 132–145.
Parment, A., & Dyhre, A. (2009). *Sustainable employer branding: Guidelines, worktools and best practices*. Malmö: Liber.
Punjaisri, K., Evanschitzky, H., & Wilson, A. (2009). Internal branding: An enabler of employees' brand-supporting behaviours. *Journal of Service Management, 20*(2), 209–226.
Rosethorn, H. (2009). *The employer brand: Keeping faith with the deal*. Aldershot: Gower Publishing Ltd.

Schivinski, B., & Dabrowski, D. (2016). The effect of social media communication on consumer perceptions of brands. *Journal of Marketing Communications, 22*(2), 189–214.
Statista. (2018). Most popular social networks worldwide as of April 2018, ranked by number of active users (in millions). https://www.statista.com/statistics/272014/global-social-networks-ranked-by-number-ofusers/. Accessed November 13, 2018.
Sullivan, J. (2002). Crafting a lofty employment brand: A costly proposition. *ER Daily*, www.erexchange.com/articles/db/. Accessed October 22, 2018.
Terjesen, S., Vinnicombe, S., & Freeman, C. (2007). Attracting generation Y graduates: Organizational attributes, likelihood to apply and sex differences. *Career Development International, 12*, 504–522.
Thomson, K., de Chernatony, L., Arganbright, L., & Khan, S. (1999). The buy-in benchmark: How staff understanding and commitment impact brand and business performance. *Journal of Marketing Management, 15*(8), 819–835.
Turban, D. B., & Greening, D. W. (1996). Corporate social performance and organizational attractiveness to prospective employees. *Academy of Management Journal, 40*(3), 658–670.
Whiting, A., & Deshpande, A. (2016). Towards greater understanding of social media marketing: A review. *The Journal of Applied Business and Economics, 18*(4), 82–91.

A Qualitative Investigation for Platform Model Conceptualization and Design: Propositions for a New Architecture

Duygu Toplu Yaşlıoğlu, Murat Yaşlıoğlu and Aykut Berber

Abstract This study aims at explaining the concept of "platform" through the network effect of multi-sided markets. Our overreaching goal is to explain how product and system designers, as platform architects, conceptualize the platform model in order to identify the platform model concept and platform strategies. In this context, to make understandable comments about platform strategy from engineer–manager perspective, a five-component model is developed using qualitative methods and focus group technique. Some propositions for platform model are:

- In order to comply with contemporary requirements and beyond-competition environment of emerging markets, companies should build platforms with acquiring new sides to their current value creation processes.
- Stimulation of strategies for current technological achievements and new requirements in the markets with better integration of data-empowered decision-making and analytics has to be exhorted.
- Structural investment and renewal for an agile company culture are necessary.
- New products and services should be designed to fulfill the requirements of multi-sided platforms and therefore form a basis for interconnected multi-agent systems to foster systematic innovations.
- Sustainability of platform base should be facilitated with right and well-informed human resource acting empowered and managing knowledge; yet as platform leaders.

Keywords Platform model · Platform design · Multi-sided networks · Platform model components

---

D. T. Yaşlıoğlu · M. Yaşlıoğlu · A. Berber (✉)
School of Business, Istanbul University, Istanbul, Turkey
e-mail: berber@istanbul.edu.tr

C. Machado and J. P. Davim (eds.), *Entrepreneurship and Organizational Innovation*, Management and Industrial Engineering, https://doi.org/10.1007/978-3-030-19289-1_4

1 Introduction

Product and system designers have the opportunities to create families of complex artifacts by developing and recombining modular components. Platforms, which are the combinations of stable components that support variety and evolution in a system, represent one of the most pronounced approaches in explaining the emergence of new strategic approaches. These platforms are observed within companies (i.e., new product development) and in higher-order inter-firm relations (collaborations). The fundamental premise behind all platforms is essentially the same; the system is partitioned into a set of "core" components with low variety and a complementary set of auxiliary components with high variety.[1] As a platform, it is subject to network effects, which tend to reinforce the previously obtained advantages such as an installed base of users, or the existence of complementary products. The emerging phenomenon of platforms affects industrial dynamics, creates new forms of competition, and reveals new forms of collaborative innovation across companies. These issues lead to new questions about the management and the design of platforms.[2] This study seeks to explain the concept of "platform" through the network effect of multi-sided markets. More specifically, our goal is to examine how product and system designers, as platform architects, conceptualize the platform model in order to identify the platform model concept and platform strategies.

Most markets with network externalities are multi-sided. Multi-sided platforms are characterized by interactions and interdependence between multiple sides (companies, departments, projects, etc.) and the existence of network effects. Accordingly, platforms devote much attention to their business model, to succeed in their industries. Industries such as software, media, payment systems, and Internet must get all sides of the market on-board. This paper emphasizes how platform models occur in different markets.

2 Concept of Platform and Platform Design

Scholars have not reached a consensus on the conceptual definition of platforms. The platform has been associated with different meanings in extant literature, such as new product development, design, and operations or technology strategy.[3] Gawer (2011) created a typology of platforms in order to organize and categorize the distinct meanings of platforms. The first typology is the internal platform which occurs inside the firm and the second typology implies the supply chain platforms which appear within a supply chain network among several firms. Third emerges in industry ecosystems, which are called industry platforms. Fourth typology namely multi-sided markets or platforms occur in industries within several firms or group of firms which are in a

[1] Baldwin and Woodard (2009), Bakos and Katsamakas (2008).

[2] Gawer (2011).

[3] Eisenmann et al. (2006): 92.

relationship with each other through the multi-sided market.[4] Platform is predominantly observed in high-tech industries. Google, Microsoft Windows, cell phone operating systems, fuel-cell automobile engines, and also some generic technologies are all examples of platforms with multi-sides that are involved.

The number and scope of inter-organizational collaborations have grown rapidly in many industries. There are two key points for collaborations to emerge, transaction, and the mutual exchange of rights. In the new business network, organizations interact more with external parties in order to access both knowledge and resources. Hence, collaboration between buyer and supplier has become a natural part of the operations of any company that develops complex products.[5] Collaborations have to ensure that three central processes are realized: (1) value creation, in order to identify a product platform for the industry, but also to evaluate this platform compared to alternative and to integrate all possible alternatives into strategic mapping process; (2) knowledge production and learning by involving partners, offering support for various experiments, and providing specific devices for knowledge production; (3) interest generation among stakeholders of the platform. To understand the fundamentals of platform architectures, one should realize the distinction between a product platform within a single firm (e.g., Sony Walkman) and a platform whose complements are supplied by many different firms (e.g., Microsoft Windows).[6]

Multi-sided networks can be found in many industries, sharing space with traditional product and service offerings. In the traditional value chain, value moves from left to right: to the left of the company is cost; to the right is the revenue. In multi-sided networks, cost and revenue are both to the left and to the right not following the traditional value chain queue, because of the platform's distinct group of users on each side. The platform is subject to costs in serving both groups and can aggregate revenue from each; albeit one side is usually subsidized.[7]

Platform businesses add value by facilitating interactions between customers who are attracted in part by network externalities. Multi-sided platforms on which it is easy to reverse participation decisions have become increasingly important since the rise of the Internet.[8] There is no perfect method or strategy for succeeding in a platform, but there are some particular steps that should be taken at the beginning. To evaluate the current state of the firm's strategies and position, management must begin with asking relevant qualitative questions. After determining the current position, management shall focus on describing the method of its market development strategies in a structured manner or otherwise criteria on the choice of a new market. With this knowledge about these factors or criteria for building a platform, it is possible to determine which choice is the right decision to build a platform and be successful

---

[4]Gawer (2011), Meyer and Lehnerd (1997), Becker and Zirpoli (2011), Gawer and Henderson (2007): 1–34, Hagiu (2009): 1011–1043.

[5]Powell (1998): 228–240, Bogers (2011): 93–117, Lindquist et al. (2008): 23–35.

[6]Gawer (2011), Moon et al. 2008, Parker and Van Alstyne (2005): 1494–1504.

[7]Eisenmann, Parker and Alstyne "Strategies for Two Sided Markets," 97; Chen (2007): 80–81.

[8]Evans and Schmalensee (2010).

in a platform.[9] If a firm wants to succeed in a platform, it should use its innovative capabilities; some of the firms which operate in industry platforms can use their core technical competences. In addition, platform architects need to make an ongoing assessment of these capabilities and the direction in which the industry is evolving. Platform architects need to focus on the platform to determine the strategic advantage of their firms over their rivals and the core competences/technologies which are difficult to imitate. A firm as a whole, in order to have a sustainable competitive advantage, should have the organizational capabilities to place itself in the center of the network of innovation through its platform.[10] When a firm wants to create or to shift to a new platform, it should keep the intellectual property of the technology and must have the right boundary units in order to monitor other companies in the market/platform.[11]

To be potential platform leader who drives the industry, a large market share most notably in an emerging market is a very good start. A successful and differentiated product makes it easier for the market leader to become the platform leader in terms of architecture, features, and technology. Open and communicative architecture allows external, complementary innovation and relationships with external developers to stimulate these innovations.[12]

Literature on platform studies is relatively new. Many recent researches have argued that a platform is not a technology, only. Platform is the outcome of a set of business behaviors and relationships between actors in the business network. Platform strategy affects product development performances, cost and lead time reduction, the international operations, and the R&D management strategies of firms. In addition, the advantages to firms pursuing a high level of platform strategy are efficiency, higher quality, and faster time to markets.[13] An early study by Muffatto (1999) reveals the benefits of adopting platform strategies through the perspective of the automobile industry where products (cars) are far more complex than those of most industries. Indicating the fact that platforms help manufacturers with flexible production, cost reduction through using resources on a world scale and thus a valuable contribution to the new product development process, the main idea of counting on such platforms when developing strategies still remains within the industry framework. However, platform strategies today tend to bring as many industries necessary to create value as possible together and therefore not to be controlled and directed by one broker company. Platform strategy affects product development performances, such as domestic and international operations and the R&D management strategies of companies. The main reasons for platform development are cost reduction, product development, and development lead time reduction. A platform strategy affects the relationship within platforms, the relationship with

[9]Davidsson and Klofsten (2003): 4–5, Sawhney (1998): 54–61.

[10]Gawer and Cusumano (2008): 68–72

[11]Gawer (2011), Sawhney et al. (2005): 4–17.

[12]Gawer and Cusumano, "Platform Leaders," 68–72

[13]Muffatto (1999a): 145–153, Koufteros et al. (2005): 97–133, Kenney and Bryan (2011).

the supplies base, subsidiaries and the other firms.[14] Another study which is about the software platform strategy suggests that there are two main issues that can introduce major challenges to introducing a platform strategy; the business strategy and product-driven platform development. Besides, firms can come across with several challenges while implementing platform strategies, such as organizational challenges (communication problems, organizational structure design, agile culture, and standardization), technical challenges (commonality and variability, design complexity, code contribution, and technical practices), people challenges (resisting to change, technical competency, and domain knowledge).[15]

Hagiu and Wright (2011) explained the drivers of multi-sided platforms in their study. Multi-sided platforms are driven by searching new sources of value and creating new network effects with the addition of new sides to multi-sided platforms. Firms need to make sure they create all the value they can deliver to their sides and with creating new network effects they can make new transactions with these sides more efficiently and more frequently.[16] The start-up associated problems are difficult to handle for firms that operate in multi-sided platforms. Firm can deliver the value from one side of the platform to the other so they have to consider and therefore balance both sides of the platform. Accordingly, firms can come across with the technological, managerial, and leadership problems while implementing their platform strategies.[17]

Relationships in the business ecosystem may take various forms. Some of these relationships are transaction-based and mostly informal using application programming interfaces (APIs) which let systems to interact with each other to carry out simpler tasks. Others are more formal and yet more complex. Contracts and service agreements are to execute governance, amplifying strategies, and so forth. Some of these partnerships may be with companies which are competitors. Alliances may be in different forms, varying from mergers to joint ventures, strategic alliances, or different layouts. As companies shuffle the right coalescence of complementary partners and allies, many forms clamping relationships that arise intricate competitive dynamics and hook players into platform systems from which it may be hard to dis-involve themselves.

Fabricating alliances force companies to scale their business and evolve. Meanwhile, these changes cannot be sustained by traditional management approaches. Leading companies find new ways to deal and form alliances with partners and rivals. An effective platform strategy strongly depends on apprehending where the value is. That rises from evaluating your boons such as customer relationships, proprietary data, capabilities, and opportunities. Companies should[18];

[14]Muffatto (1999b): 449–460, Ulrich (1995), Clark and Fujimoto (1991), Muffatto and Roveda (2002).

[15]Ghanam et al. (2012).

[16]Hagiu and Wright (2011): 1–7, Hagiu (2007).

[17]Gawer (2011), Economides and Katsamakas (2006).

[18]Meffert and Swaminathan (2017), Lee (2013): 66–110.

1. Evaluate the ***market*** in which your potential partner operates and its ***level of competition***. The most promising ecosystems involve market leaders with complementary skill sets and value propositions.
2. Consider the company's ***business model***. Is it fit for purpose and future-proof? What products and services does the company produce? How nimble, innovative, and customer-focused is it? Can it keep pace with you and the external environment?
3. Weigh the ***human factor***. How strong is the company's management team? How effective are its employees?
4. Look at the ***culture***. How does your potential partner do business? How does its way of working fit with your own company's culture?

Some platform archetypes, their objectives, and examples are given in Exhibit 1.

3 Research Question and Methodology

The framework presented above provides a starting point for this research. This framework is examined by the help of a study themed "implementation of building a successful platform model in multi-sided markets," within a focus group. As indicated earlier, the primary purpose of this study is to explain how product and system designers, as platform architects, conceptualize the platform model in order to identify the platform agents/sides and platform strategies. To obtain information concerning "platform model design" and "how product and system designers can achieve to be a platform architect in order to build a successful platform business model", engineers as platform architects from different hierarchical levels of companies were interviewed. In this focus group case study, sixteen engineers (in two separate groups eight each) who were, at the time being, taking graduate degree "management and organization" course were interviewed. People who were included in the interview graduated from machinery, metallurgy, chemistry, industrial, food, and forestry engineering faculties. They were working in different sectors and different departments such as marketing, business development, and quality management. Since mentioned above, the constitution of platforms aims to provide value creation, knowledge production, and creating value at the industry level. This various engineering background combined with the education of management and organization field gave us the very best subjects as platform architects. Questions were asked to get interviewees' ideas and thoughts about the platform model. These people were chosen because they are working as product and system designers and because of their role as a platform architect at their job. Interviewees' various backgrounds led us to obtain information from different perspectives. These groups' interview averaged about 75 min. To improve the internal validity of the data, interview questions were constructed from a variety of information sources. During the interview period, for the quality and comprehensibility of the interview, we (also the moderator of the focus group) supplied interviewees with mini-cases about the subject

	Objectives	Examples
Sales	• Allow users to interact with each other socially • Primary angle is to attract as many customers and social interactions as possible	• Individual social platforms such as Facebook, Tencent Wechat, Instagram, Twitter, Line • Professional social platforms such as LinkedIn
Data	• Sharing data using standard data definitions • Providing additional data-and analytics-based services • Primary focus is to capture and use external and proprietary data	• IOT efforts by John Deere, Cater pillar Minestar, and GE Aviation • Apple iHealth • P&C insurers (e.g. using weather data patterns to assess fire risks)
Technology	• Enablement by industry standard software and hardware • Better/faster IT delivery through broad range of specialist IT firms and technologies	• Consumer mobile such as Apple iOS and Google Android • Enterprise technology platforms such as SAP ERP, Microsoft Office, Oracle ERP, SAP Hans • Visa/MasterCard's payment processing platforms, and Bitcoin
Customer journeys	• Leverage company's core commerce functionality and value proposition to attract large number of customers • Additional capabilities to complete the customer journey and experience can then be added to create network effect	• Ride sharing platforms such as UBER and Lyft • Shopping platforms such as Amazon • Travel platforms such as Airbnb • Banks allying with FinTech players in value chain, e.g., SME app players linked via APIs into bank
Services	• Integrate multiple companies' services together to holistically address customers' pain points and make the initial product/services much more attractive for customers • Bundling can be done either transparently or explicitly	• Transparent add-ons such as Slack and Amazon Alexa • Explicit services platform such as Salesforce.com and the salesforce ecosystem/AppExchange

Exhibit. 1 Platform Archetypes. *Source* Meffert and Swaminathan (2017)

and showed pictures and diagrams about multi-sided market examples. Interviews with the managers of the future of platform businesses also expand the scope of the research and give an opportunity to examine the impact of the academic background (maturity) of engineers on this framework. By this way, the in-depth investigation of this business model not only will help us to evaluate the framework, but also help to collect the participants' opinions on multi-sided markets and platform business model implementation process.

Interview questions were designed based on previous studies and questionnaires from the literature. Interview questions were designed to provide guidance, to encourage consistency, and to allow respondents more freedom of expression to learn more about platform model implementation in a business environment. Interviews also consist of open-ended question to extract more data. To learn the most important perspective about the implementation of platform model, all interviews started with these questions:

- Under which conditions can we expect industry platforms to emerge?
- What forms do they take? In what context?
- How can firms succeed in the new platform game?
- Which capabilities are needed?
- When should firms open up their technologies and processes?
- Which strategies can a firm implement to adopt their firm a platform model?
- How to manage the exploratory processes to create a platform?
- How to manage platform development projects?
- What drives multi-sided platforms?
- Which problems will occur while implementing platform strategy?

4 Findings

Insights from our interviews were digitally recorded. The following step was the verbatim-translation and examination of interview findings. Afterward, these listed but untouched answers were grouped; the answers which were considered to have the same/close meaning were either omitted or converged. However, while purifying the answers, the count of mentions was crucial. Since the main purpose was to categorize the concept, most mentioned and therefore most valued answers were in our main concern. While counting the mentions of any idea or opinion, taking into account that one could have mentioned his/her idea more than once and/or implied the same meaning without using the exact word, converging and yet unification into solitary meaning was the foremost important and delicate stage of this process.

Following the purification and deduction process, going through the results of the interviews and classifying them, we constituted out five factors, which are later called "Platform model design factors." These factors are: stimulating (technological development, new requirements in the sector, etc.), structural (making useful strategic plans, acquire innovative people, etc.), design-related (platform-based product design, developing a multi-agent system, etc.), sustainability-related (searching the right people to emerge platform strategies, difficulty in determining the target audience, etc.) and managerial (using social media, accurate market research through the right and well-informed people, creating a team to implement platform strategy etc.) factors. Further and yet separate step for this research is planned to categorize these extracted concepts into variables, and test them with quantitative analyses. This latter step was only possible with categorizing the platform concept; therefore, this research is the most crucial step for a quantitative model formation.

Some characteristics are mandatory for a market or network to qualify as a multi-sided platform. This paper attempts to contribute toward building, maintaining, and developing platform model. One of the main conclusions emerging from our analysis is identifying a platform's new "side" (or new "sides") could create strong network effect within an existing firm. Also, a firm's relationship with customers and suppliers is very important for designing platform concept. In order to achieve platform strategy, first there shall be a manager called "platform leader/architect" and second should there be his/her project team. This team can face various problems while composing and implementing platform design applications. Main characteristics of the platform model and its features are given above.

Platform design factors, issues, and characteristics aforementioned are limited to this study's sample and cannot be generalized without further research. But this paper gives researchers and academicians a general idea about implementing and developing a platform model and multi-sided networks. Moreover, this paper contributes to literature as a starting point for further research.

In order to make understandable comments about platform strategy from engineer–manager perspective, the components of the platform model are analyzed and constituted as such:

1. Stimulating characteristics: Developing technology, emerging "new sides" and "new requirements" in the market, foreseeing the future, and making a difference in the product.
2. Structural characteristics: Flexible, acquiring innovative people, understanding the customers' requests, making good strategic plans and having a responsive structure that able to respond to customers' needs quickly, and giving importance to suppliers and agents which form the platform.
3. Design characteristics: Extending and extracting concepts from platform-based product design, developing a multi-agent system to support platform design, applying agent-based decision making to generate and determine platform design strategies.
4. Difficulties: "Platform agents are the basis"; Need for constantly developing the system, need for removing deficiencies, difficulty in determining and reaching the target market segments, difficulty in following rivals' strategies, and problems in finding right people to emerge and diffuse platform strategies toward the customers and the market.
5. Behavioral characteristics of the platform manager/leader: "There is a need for managers who manage the knowledge." Providing the right feedback to/from all

agents of the platform, making network analysis of the platform and resolving the network, using social media efficiently and effectively, and accurate market research through the correct and well-informed people.

Management characteristics of the platform strategy projects: "Design of the product is not solely sufficient." Making preliminary study or pilot research about implementing the platform strategy (first he/she can begin with a department of an agent or specific products), creating a team of "right" members to work with, using technical knowledge of engineers, being open to new ideas, developing good relationships with suppliers as an agents of platform, maintaining the highest level of quality standards, and identifying a platform along with variant and unique modules using data mining techniques.

5 Discussion

According to the literature, some main characteristics are necessary for a market or network to qualify as multi-sided platforms. First, two or more categories of agents must exist; second, an agent of one category should benefit in a way from the presence of members from other categories. Third, agents shall not internalize these benefits, which can make positive profits by connecting the two types of agents in a more efficient way than what they could achieve on their own.[19] In our research, we found that, developing technology, foreseeing the future, and considering new sides in the market are stimulating characteristics of a platform model. In relation to this, if a firm wants to develop a platform model, it should have a good strategic plan, innovative people, and flexible organization culture.

If a firm wants to succeed in a platform, it should use its innovative capabilities; so it should have more innovative people because these people are going to be platform architects in the future. In our research, we figure out that design is not sufficient alone, so platform architects need to focus on the platform to determine the strategic advantage of their firms over their rivals and the core competencies which are difficult to imitate, it is parallel with the literature.

A platform strategy affects the relationship within platforms, the relationship with the supplies base, subsidiaries and the other firms, hence platform leaders or managers of a firm come across with several difficulties while implementing a platform strategy, such as difficulty in determining and reaching the target market segments, difficulty in following rivals' strategies, and so on.

Most of our conclusions are parallel with the literature but we figure out some other key points like, there are too many different characteristics of a platform and we classify them into three main components: stimulating, structural, and design characteristics. There is no other study in the literature that have classification like these, also we classify management team's and platform leaders' characteristics so we found that there should be different behavioral styles between them.

---

[19]Hagiu, "Merchant of Two-Sided Platforms", 5–11.

The two distinct user groups interact with each other through a common network platform; these network platforms coordinate the demands of the two sides of the market that need each other and help them to make money by linking different sides of their customer networks. They provide infrastructure and transaction rules that facilitate different sides' interaction and transaction. As an example, computer operating system vendors such as Microsoft provide software on which computer users and application developers can be linked together; credit card companies provide plastic cards and authorization terminals through which merchants and consumers can make transactions. Platforms can be physical or virtual places providing services, such as real estate brokerage, shopping malls, and Web sites like Monster and eBay.[20]

The platform model design issues on which we have focused on this paper is constituted from only "ideas of engineers" about being a platform model business concept. This paper does not give insight about pricing and distribution of industrial problems of a specific sector or a firm. It was aimed to draw general ideas about implementing and developing a platform model and multi-sided networks, and according to these ideas, more specific researches can be made for a further study.

6 Propositions

In order to clarify the concept of "platform" and the formation of platform structures, we have several propositions which should be applied overall enterprise. These extracted propositions can also be put in a modal in later researches to create a better and more concrete foundation for platform strategy.

Proposition 1 *In order to comply with contemporary requirements and beyond-competition environment of emerging markets, companies should build platforms with acquiring new sides to their current value creation processes.*

The driving force behind multi-sided markets is the need to induce coordination among two or more groups of agents, and what they "coordinate on" is precisely a fixed point in the architecture of transactions in which they collectively participate. That fixed architectural point may be a particular component of system; such as Visa payment-processing system which both issues cards to consumers and approves transactions on behalf of merchants; or a physical location, as in the case of a shopping market which provides meeting point for both merchants and consumers. Moreover, it may also be a convention, such as the use of currency as a medium of exchange, or compatibility between systems. One of the main conclusions emerging from our analysis is that, identifying a new side (or new sides) that could create strong network effect with the existing firm. Also firm's relationship with customers and supplies is very important for designing platform concept.

[20]Chen (2007): 80–81.

Proposition 2 *Stimulation of strategies for current technological achievements and new requirements in the markets with better integration of data empowered decision making and analytics has to be exhorted.*

Data is raw information, which is processed into knowledge. Relentless and ever-ending collection of data from all over the internet search engines, mobile phones, broadcasts, applications, web sites, social media, blogs, wikis and so on, has created an opportunity and a great deal of knowledge source in the form of raw data. However, this data has piled up unimaginably after the digital revolution and recently this repository has increased its pace more than an exponential function. This vast repository can also be harnessed by analytics, and therefore eventually used for decision making in various areas. Traditional sources of data such as reports, excel files, SQL were the tools of traditional managers, or as mentioned before "ITsmiths" and information workers; contemporary managers or so-called "digital leaders" need and harness their information from analytics tools and depend on a real-time source of information. This, hence, gives less chance of success to traditional managers. Such digital competence is not just a key for success but it must be considered as a bottom-line for knowledge created throughout. A platform architect or leaders therefore needs to have such skills and education to perform aforementioned tasks; moreover, the employees must conform with similar skills.

Proposition 3 *Structural investment and renewal for an agile company culture is necessary.*

Enterprises with the highest degree of agility had more services in each cluster and broader implementations of each service. Strategic agility requires time, money, leadership and focus—and an understanding of which distinct patterns of high-capability infrastructures are needed where.[21] *Agile culture is formed through and contributed with learning orientation, skilled knowledge workers, collaboration and communication, shared vision and balanced allocation of resources.*[22] *Learning orientation refers to capability refers to the ability of organizations to learn, learn how they learn and renew themselves in a timely manner.*[23] *Resource Allocation Planning includes forecasting and decision making about how to allocate resources for strategy execution. Such resources include the people, time and budget. Doz and Kosonen*[24] *describe several organizational conditions that influence the Resource Allocation Planning process. These include the pace of such allocations, collaboration and inclusion, maintaining a systems view, and perceived fairness as central to the effectiveness of resource planning. They relate perceived fairness of the process to the ensuring the process is open and transparent and the decision-making is based on shared values of collaboration and system thinking of the good for the entire company.*

[21] Weill et al. (2002): 57.

[22] Adler (2014) San Francisco.

[23] Argyris (1999), Senge (1998).

[24] Doz and Kosonen (2008).

Proposition 4 *New products and services should be designed to fulfil the requirements of multi-sided platforms and therefore form a basis for interconnected multi-agent systems to foster systematic innovations.*

Systematic Innovation for Business focuses the problem solver on the root problem to be solved and breaks the mental inertia around how that classification of problem has been traditionally solved. Further, Systematic Innovation for Business identifies and helps to negate conflicts facing your organization. The result is a clear understanding of what issues need analysis focus and will best address the organization's challenges in a holistic way.[25] *Innovation used to be a linear trajectory from new knowledge to new product. Now innovation is neither singular nor linear, but systemic. It arises from complex interactions between many individuals, organizations and their operating environment.*[26] *Systematic innovation needs a cross tabulation of ideas to create a package of new products or services. One example of systematic innovation is e-automotive industry. It is impossible to think of electric cars without adequate batteries to charge and charging stations built for juice them up. Hence, a good network for knowledge transfer and sharing should be based among multi-agent based technology and service creators.*

Proposition 5 *Sustainability of platform base should be facilitated with right and well informed human resource acting empowered and managing knowledge; yet as platform leaders.*

In order to achieve platform strategy, there should be a manager called "platform leader/architect" and there should also be a project team with suitable members to platform design. This team can face many difficulties while composing and implementing strategy related applications. Outlining characteristics about the platform model and its features are discussed in findings section of this research. These characteristics must be considered while building and sustaining so called platforms which are in essence multi-sided networks.

Platform architects have to have exceptional strategic skills and collaborative capabilities, they need to be skilled knowledge workers whom harness the data and turn it into knowledge. Skilled knowledge workers imply the human resources to align its planning, selection, evaluation and education functions to the level of organizational maturity. Vision is the shared dedication to the organization's goals by the leadership and employees across the organization.[27] *Besides platform architects a serious implementation of empowerment in the system to create multiple self-managed platform leaders is a must. Therefore, traditional task oriented approach for sides of the platforms is insufficient, a more valorous skill based approach where employees have the competencies to cross-functionalize the multi sides of platforms. Since the concept is latent enough to vanish by itself, the awareness of such communication and collaboration channels have to be sustained by knowledge workers or otherwise platform and sub-platform leaders. First is to continuously assess what*

---

[25] InnomationCorp (2018).

[26] 1000Ventures (2018).

[27] Beer and Eisenstat (2000): 29, Doz and Kosonen (2008).

skills and innovations are needed for organization goal achievement and successful strategy execution and to develop them in a timely manner. Second, is to reflect in an effective manner on the entire process that is occurring in the first three capabilities (the assessment, direction and execution) and make adjustments as needed. Multi-sides' differences and requirements should have been reconciled with more dialogue and attempts to collaborate and communicate in line with vision. Therefore, a basis of communication and collaboration should be established throughout the entire platform.

References

1000Ventures. Systematic Innovation. http://www.1000ventures.com/business_guide/innovation_systemic.html. Accessed June 2018.

Adler, N. (2014). *The strategically agile organization: Development of a measurement instrument* (Ph.D. dissertation). Alliant International University, California School of Professional Psychology, San Francisco.

Argyris, C. (1999). *On organizational learning*. Malden, MA: Blackwell Publishers Ltd.

Bakos, Y., & Katsamakas, E. (2008). Design and ownership of two-sided networks: Implications for internet platforms. *Journal of Management Information Systems, 25*(2), 171–202.

Baldwin, C. Y., & Woodard, C. J. (2009). The architecture of platforms: A unified view. *Platforms, Markets and Innovation*, 19–44.

Becker, M. C., & Zirpoli, F. (2011). *Beyond product architecture: Division of labour and competence accumulation in complex product development*. Department of Management, Universita Ca' Foscari Venezia, Working Paper no. 3.

Beer, M., & Eisenstat, A. R. (2000). The silent killers of strategy implementation and learning. *MIT Sloan Management Review, 41*(4).

Bogers, M. (2011). The open innovation paradox: Knowledge sharing and protection in R&D collaborations. *European Journal of Innovation Management, 14*(1).

Chen, K. (2007). *Dynamic competition of two-sided platforms: Differentiation, pricing and strategies* (Ph.D. dissertation). Stanford University, 2007.

Clark, K., & Fujimoto, T. (1991). *Product development performance: Strategy, organization, and management in the world auto industry*. Harvard Business Press.

Davidsson, P., & Klofsten, M. (2003). The business platform: Developing an instrument to gauge and to assist the development of young firms. *Journal of Small Business Management, 41*(1).

Doz, Y., & Kosonen, M. (2008). How agile is your strategy process. *Strategy Magazine*, 15(1), 6–10.

Economides, N., & Katsamakas, E. (2006). Two-sided competition of proprietary vs. open source technology platforms and the implications for the software industry. *Management Science, 52*(7), 1057–1071.

Eisenmann, T., Parker, G., & Van Alstyne, M. W. (2006). Strategies for two-sided markets. *Harvard Business Review*, 84(10).

Evans, D. S., & Schmalensee, R. (2010). Failure to launch: Critical mass in platform businesses. *Review of Network Economics*, *9*(4).

Gawer, A. (2011). *Platforms, markets and innovation*. Edward Elgar Publishing.

Gawer, A., & Cusumano, M. A. (2008). Platform leaders. *MIT Sloan management Review*.

Gawer, A., & Henderson, R. (2007). Platform owner entry and innovation in complementary markets: Evidence from intel. *Journal of Economics & Management Strategy, 16*(1).

Ghanam, Y., Maurer, F., & Abrahamsson, P. (2012). Making the leap to a software platform strategy: Issues and challenges. *Information and Software Technology, 54*(9).

Hagiu, A. (2007). Merchant or two-sided platform? *Review of Network Economics, 6*(2).
Hagiu, A. (2009). Two-sided platforms: Product variety and pricing structures. *Journal of Economics & Management Strategy, 18*(4).
Hagiu, A., & Wright, J. (2011). Multi-sided platforms. Boston: Harvard Business School. MA Working paper 12-024.
Innomation Corp. What is Systematic Innovation? http://innomationcorp.com/si_defined.html. Accessed June 2018.
Kenney, M., & Bryan, P. (2011). Structuring the smartphone industry: Is the mobile internet OS platform the key? *Journal of Industry, Competition and Trade, 11*(3), 239–261.
Koufteros, X., Vonderembse, M., & Jayaram, J. (2005). Internal and external integration for product development: The contingency effects of uncertainty, equivocality, and platform strategy. *Decision Sciences, 36*(1).
Lee, R. S.-J. *Essays on platform competition and two-sided markets* (Ph.D. dissertation). Harvard University 2013.
Lindquist, A., Berglund, F., & Johannesson, H. (2008). Supplier integration and communication strategies in collaborative platform development. *Concurrent Engineering, 16*(1), 23–35.
Meffert, J., & Swaminathan, A. (2017). Management's next frontier: Making the most of the ecosystem economy. *McKinsey Insights*. https://www.mckinsey.com/business-functions/digital-mckinsey/our-insights/managements-next-frontier. Accessed June 2018.
Meyer, M. H., & Lehnerd, A. P. (1997). *The power of product platforms: Creating and sustaining robust corporations*. New York: Simon and Schuster.
Moon, S. K., Simpson, T. W., & Kumara, S. R. T. (2008). A strategic module-based platform design method for developing customized products in dynamic and uncertain market environments. In *ASME 2008 International Design Engineering Technical Conferences and Computers and Information in Engineering Conference*. American Society of Mechanical Engineers, 2008.
Muffatto, M. (1999a). Introducing a platform strategy in product development. *International Journal of Production Economics, 60*.
Muffatto, M. (1999b). Platform strategies in international new product development. *International Journal of Operations & Production Management, 19*(5/6).
Muffatto, M., & Roveda, M. (2002). Product architecture and platforms: A conceptual framework. *International Journal of Technology Management, 24*(1), 1–16.
Parker, G. G., & Van Alstyne, M. W. (2005). Two-sided network effects: A theory of information product design. *Management Science, 51*(10), 1494–1504.
Powell, W. W. (1998). Learning from collaboration: Knowledge and networks in the biotechnology and pharmaceutical industries. *California Management Review, 40*(3), 228–240.
Sawhney, M. S. (1998). Leveraged high-variety strategies: From portfolio thinking to platform thinking. *Journal of the Academy of Marketing Science, 26*(1), 54–61.
Sawhney, M., Verona, G., & Prandelli, E. (2005). Collaborating to create: The internet as a platform for customer engagement in product innovation. *Journal of Interactive Marketing, 19*(4), 4–17.
Senge, P. M. (1998). The leader's new work. *Leading Organizations, 1*, 439–457.
Ulrich, K. (1995). The role of product architecture in the manufacturing firm. *Research Policy, 24*(3), 419–440.
Weill, P., Subramani, M., & Broadbent, M. (2002). IT infrastructure for strategic agility. *MIT Sloan Management Review, 44*(1).

The Utility of Human Resource Managers' Action: A Self-centred Perception by Different Organizational Actors

João Leite Ribeiro, Delfina Gomes and Ana Caria

Abstract This study analyses how different organizational actors perceive and characterize human resource (HR) managers' actions at the strategic, tactical and operational levels. In this study, 257 interviews were conducted with organizational actors, and Grounded Theory was applied to analyse the data. The interviewees were employed at ten companies (three multinational and seven Portuguese companies) and represented different hierarchical positions: top managers, peers of HR managers and collaborators from different organizational functions. The interviewees expressed very clear views regarding desirable axes of action for HR managers, and these actions could improve perceptions of HR managers and departments. These desirable axes of action were as follows: proactive; strategic combined with an operational level of action; demonstrating initiative; pragmatic; simultaneously global and local (*glocal*); transparent; objective; just and impartial; legitimate; and efficient and effective. Therefore, the results highlight three main axes where HR managers need to focus their attention: principles of practical action, principles of communication and ethical values. The results demonstrate that when developing their identities and credibility within organizations, HR managers must consider the perceptions of different organizational actors. The ways in which HR managers are perceived by different organizational actors pose a challenge and have an effect on the roles and influence that HR managers and departments can have at the strategic level in organizations. This study contributes to the literature by exploring an important yet underdeveloped field of research on how HR managers are viewed within organizations.

Keywords Human resources management · Human resources managers · Axes of action · Perception · Grounded theory

---

J. L. Ribeiro (✉) · D. Gomes · A. Caria
School of Economics and Management, University of Minho, Braga, Portugal
e-mail: joser@eeg.uminho.pt

C. Machado and J. P. Davim (eds.), *Entrepreneurship and Organizational Innovation*, Management and Industrial Engineering, https://doi.org/10.1007/978-3-030-19289-1_5

1 Introduction

HR strategic management presupposes an interconnection with an organization's business and results from alignment between HR management strategies, business strategies and the organization's overall strategies (Bloom et al. 2012; Buyens and De Vos 2001). This alignment should be based on reliable human resources management (HRM) policies and practices that are consistent with those of other internal areas of the organization. Additionally, there must be an equal level of coherence and consistency between the same policies and practices (Aldrich et al. 2015; Boswell 2006; Brewster 1997; Guest 1987, 2010; Guest et al. 2003; Hendry and Pettigrew 1990; Ibănescu 2015; Markoulli et al. 2017; Miles and Snow 1984; Rivero and Dabos 2017; Schuler and Jackson 1987; Stone and Deadrick 2015; Waldman et al. 2012).

In the strategic perspective of HRM, certain approaches assume a more reactive approach through which HR programmes, processes and practices contribute to the achievement of an organization's objectives (Gannon et al. 2015; Miles and Snow 1984; Stavrou and Brewster 2005). HR managers may participate in the strategic planning process through the use of various proactive approaches, and the available HR or characteristics that they have may condition the organization's strategies (Gannon et al. 2015; Miles and Snow 1984; Rivero and Dabos 2017). In daily practice, the routinization of many HRM responsibilities occurs, although it is expected that HR managers and departments will assume a proactive stance and a strategic position within their organizations (Rivero and Dabos 2017; Tung 2016).

This poses new challenges for HR managers, who must reevaluate the way they position their work role (e.g. in legal terms), redefine their goals, values, and roles and develop new competencies and work practices (Cabral-Cardoso 2004; Rivero and Dabos 2017; Stone and Deadrick 2015; Tung 2016; Ulrich and Dulebohn 2015). However, obstacles to the development of HRM must be considered, namely the lack of skills and motivation among many HR managers to effectively play a strategic role in the organizations in which they work (Bloom et al. 2012; Cabral-Cardoso 2004; Scullion and Starkey 2000; Torrington 1998). In addition to these obstacles, top managers sometimes use a discourse through which they advocate the need for a strategic and proactive HR function even though there is no real understanding of what this means. This lack of understanding has implications for specific management measures that may be implemented in an organization by HR managers and departments (Hutchinson and Purcell 2010; Lo et al. 2015; Mirfakhar et al. 2018; Ng 2011; Perry and Kulik 2008; Uyar and Deniz 2012). Other obstacles have been highlighted, such as ambiguities and inconsistencies at various levels of HRM and notably in terms of the nature and exercising of the authority of HR managers (Bloom et al. 2012; Cabral-Cardoso 2004; Davila et al. 2010; Torrington 1998). These obstacles are incremented by the fact that HR departments have to accommodate and serve institutional, instrumental and individual objectives, simultaneously, and serve multiple stakeholders or constituencies (Tsui 1990; Tsui and Milkovich 1987).

This paper investigates the perceptions of the activities and responsibilities of HR managers among different organizational actors at different hierarchical and functional levels within the organization in the Portuguese context. This focus is related to the role that HR managers play in organizational strategies as perceived by other actors, with a particular emphasis on the different levels of action of HR managers; these different levels are organized according to categories termed axes of action. Therefore, we aim to better understand how different organizational actors perceive and characterize HR managers' actions at the strategic, tactical and operational levels. This study addressed the following research question: What are the current perceptions of HR managers and how are these perceptions affected by the axes of action perceived by different organizational actors?

This empirical study used interviews to collect data; 257 interviews were conducted with different organizational actors, and Grounded Theory was used to analyse our data. The interviewees were employed at ten companies (three multinational and seven Portuguese companies) and were at different hierarchical levels: top managers, peers HR managers and collaborators from different organizational functions. These companies are mainly large companies and leaders in their fields of business and were selected only companies with an HR department and an HR manager who is hierarchically and/or functionally subordinate to an administrator, general manager or superior general HR manager. Although not representative of the Portuguese business structure, composed mainly by small and medium companies, it is important to consider the Portuguese context of HRM. HRM has developed from a purely bureaucratic and administrative role to a people-centred approach to management (Cabral-Cardoso 2004). However, the function of HR manager is a reality mainly in medium and large companies. According to the Cranet survey, Portugal was considered to possess a lower level of strategic integration, but with line managers playing a significant role in HRM (Cabral-Cardoso 2004).

This study contributes to the literature by exploring an important yet underdeveloped field of research on how HR managers are perceived within organizations (Brandl and Pohler 2010; Farndale 2005; Uyar and Deniz 2012; Yamamoto 2013). This study also contributes to the debate on how the perceptions of other organizational members have an effect on the roles that HR professionals play at the strategic level in organizations and, consequently, on the success of HRM practices (Aldrich et al. 2015; Mirfakhar et al. 2018; Sheehan et al. 2014). This study also contributes to practitioners, by highlighting that HR managers and departments serve multiple stakeholders or constituencies, with multiple and diverse objectives (Tsui 1990). As stated by Ulrich et al. (2017, p. 10), "understanding stakeholder expectations defines who HR must satisfy to help the firm succeed". Therefore, the results highlight three main areas of action where HR managers need to focus their attention. First, the principles of practical action, which underpin their practices on HRM. Second, principles of communication, which provide the guidance for a clear and effective communication with the stakeholders of HRM. Third, the ethical values which are the basis both for their practices of HRM and for their interaction with HRM stakeholders.

The remainder of this paper is structured as follows. The next section presents a brief outline of the literature on HRM and on HR managers from the strategic perspective. The methodology of the present study, specifically the methods of data collection and analysis used, and a description of our participants are presented in the section following. Next, the perceptions of different organizational actors regarding the axes of action of HR managers are described and analysed. There follows the discussion of the results and the theoretical model. The final section presents the conclusion, limitations and future research.

2 HRM and HR Managers

An understanding of the perceptions of HR managers and departments is a topic that continues to attract the attention of researchers (Brandl and Pohler 2010; Farndale 2005; Stavrou and Brewster 2005; Uyar and Deniz 2012; Yamamoto 2013).

Different authors have highlighted that the analyses of HR departments should consider several factors, such as the approach or methods used to manage this department and existing HR development policies and strategies for identifying and recruiting new talent (Stone and Deadrick 2015; Yamamoto 2009). Such research has also emphasized how HR departments are built, formalized and institutionalized and how their image and expectations of their performance are developed (Coda et al. 2009; Yamamoto 2013). Therefore, the study of HR managers and departments, as well as the implementation of HR practices, needs to consider the conditions created by macro-, mezzo- and micro-contextual elements, since the actions taken by different organizational members do not occur in a vacuum (Cooke and Saini 2010; Mirfakhar et al. 2018; Tsui 1990). Important is also the fact that the HR department has to engage and serve multiple stakeholders (Klaas et al. 2012; Mirfakhar et al. 2018; Tsui 1990). As argued by Tsui and Milkovich (1987), the ability of HR departments to meet these stakeholders' expectations or demands may dictate its survival.

2.1 *The Nature of Specific HR Departments*

An organization's HR department and its importance can be analysed, on the one hand, from the perspective of the formulation of principles, policy development, guidelines and actions, and the design and implementation of HRM practices (Bloom et al. 2012; Dolan et al. 2010; Lo et al. 2015; Mirfakhar et al. 2018; Pernkopf-Konhäusner and Brandl 2011; Yamamoto 2009). On the other hand, it is possible to analyse these bodies based on the power that they assume or can assume within an organization (Bloom et al. 2012; Hennessy and McCartney 2008; Mirfakhar et al. 2018; Stainback et al. 2010). In addition, the characterization of HR departments can be carried out based on strategic and operational dimensions (Dolan et al. 2010; Farndale 2005; Hennessy and McCartney 2008). From the operational point of view, this characterization is structured in terms of the type of policies, programmes and practices that are conceptualized, developed, operationalized and implemented by these departments. From the strategic point of view, this characterization is carried out through the role that these departments assume throughout the whole process of planning and strategically formulating nuclear principles, from the vision to the organizational mission. However, it is important to take into consideration that the HR department interacts with a variety of constituencies, including top management, line managers and employees of other functional departments, with multiple objectives and potentially inconsistent (Tsui 1990).

As proposed by Ulrich (1997a, b, 1998a, b), the role of HR departments can also be analysed based on role typology. This typology is based on the roles assumed and developed by the people and on the different HR processes, on the one hand, and on the operational and strategic activities that the HR body develops, on the other. More importantly, HR departments need to create a comprehensive information advantage and deliver measurable business value (Ulrich et al. 2017) to help organizations overcome the different challenges and demands they face (Boon et al. 2011; Truss et al. 2002; Yamamoto 2013).

The relevance of this department, besides its strategic and operational relevance, is also tactical and instrumental, based on multiple factors. These factors include the structure of an organization, the composition and location of a department in an organization's chart, and its nature, scope and level of assigned functions and responsibilities (Aldrich et al. 2015; Dolan et al. 2010; Pfeffer 1998). However, several studies have highlighted that many HR departments are still far from being involved with strategy issues and from contributing to their organization performance (Brandl and Pohler 2010; Farndale 2005).

As with the research on HRM and HR managers, the literature on HR departments has also used two distinct approaches: what they do and what they should be doing. Thus, this study addresses a very important dimension: expectations surrounding the roles and axes of action of HR managers and HR departments. Considering the transformations that have occurred in HRM and the roles and respective expectations of HR managers and departments, key facilitators and inhibitors that may facilitate a better understanding of these changes are discussed below.

2.2 Facilitators and Inhibitors of the Roles of HR Managers and Departments

The role and status of HRM in an organization, while relying largely on factors internal to the organization, are also dependent on external factors. This is attributed to the fact that any organization as a system interacts with the external environment and is influenced by this same context (Bloom et al. 2012; Stainback et al. 2010; Ulrich 1997a, b, 1998a, b). In fact, the context and its characteristics are paramount in explaining the success or failure of HRM practices (Mirfakhar et al. 2018).

In internal terms, the effective importance that senior managers attribute to HRM is of particular relevance. This importance allows HRM to assume a strategic function within an organization. This strategic function can help HRM actively and proactively anticipate and solve problems while adding value and creating competitive and distinctive organizational advantages in the short, medium and long terms (Aldrich et al. 2015; Bloom et al. 2012; Cabral-Cardoso 2004; Farndale 2005; Paauwe and Boselie 2003; Waldman et al. 2012; Yamamoto 2013). Another important factor related to the success of organizations and particularly of business concerns the ability to identify, to support and to optimize people's potential and competencies (Bloom et al. 2012; Boselie and Paauwe 2005; Cabral-Cardoso 2004; Mirfakhar et al. 2018; Paauwe and Boselie 2003; Pfeffer 1995; Sheehan et al. 2014; Yamamoto 2013). Also relevant is the capacity of HRM to keep a positive relationship and high-quality communication with HR stakeholders (Klaas et al. 2012; Mirfakhar et al. 2018).

However, obstacles to the development of HRM must be considered, and they may emerge from specific characteristics of different actors. For example, the lack of skills and motivation among many HR managers to effectively play a strategic role in the organizations in which they work (Bloom et al. 2012; Cabral-Cardoso 2004; Scullion and Starkey 2000; Torrington 1998). Other obstacles have been highlighted, such as ambiguities and inconsistencies at various levels of HRM and notably in terms of the nature and exercising of the authority of HR managers (Bloom et al. 2012; Davila et al. 2010; Mirfakhar et al. 2018; Torrington 1998). Other important actors are top managers, who sometimes use a discourse through which they advocate the need for a strategic and proactive HR function even though there is no real understanding of what this means. This lack of understanding has implications for specific management measures that may be implemented in an organization by HR managers and departments (Hutchinson and Purcell 2010; Lo et al. 2015; Mirfakhar et al. 2018; Ng 2011; Perry and Kulik 2008; Uyar and Deniz 2012). Also, line managers' skills, beliefs and attitudes, as well as the support they receive from the organization in developing their work, play a significant role, positive or negative, in the development and implementation of HRM practices (Dewettinck and Vroonen 2017; Mirfakhar et al. 2018). Finally, employees' characteristics, such as ability to organize their work to meet their personal goals and previous experiences with HRM, play a significant role as facilitators or as obstacles in the development and implementation of HRM (Meijerink et al. 2016; Mirfakhar et al. 2018; Piening et al. 2014).

In the specific Portuguese context, and in what concerns HR managers, the studies argue that they are still struggling to be listened in the organization and to set the HRM agenda (Cabral-Cardoso 2004). Two additional difficulties to the consolidation of the HRM in Portugal are the lack of credibility of the HR function and the size, ownership and structure of companies (Cabral-Cardoso 2004). The Portuguese HRM context, as described by Cabral-Cardoso (2004, p. 973; see also House et al. 2004), "is sometimes described as one in which 'organization' is seen as an abstract and far away entity most employees would not identify themselves with, thus hindering the development of affective organizational commitment".

Next, we describe methodological and research methods used for the present study.

3 Methodology[1]

This study is part of a broader investigation developed under an interpretative paradigm by assuming that reality is a social construction and cannot be understood independently from the actors that create that reality (Urquhart 2013). The empirical study is based on qualitative research designed to understand phenomena through the meanings that individuals attribute to them (Myers 2011). This study used interviews for data collection purposes. Grounded Theory methodology was adopted to analyse the interviews (Corbin and Strauss 2008; Laperrière 2010; Strauss and Corbin 1998; Urquhart 2013). This analysis was confirmed and validated by a specialist in the use of Grounded Theory.

The data collected through the interviews reveals a set of categories that were grouped into broader dominions, which we designate as *axes of action*, to organize the coding process. These axes[2] of action derived from the data convey, from a deeper analysis of the discourse produced, continuums of guiding courses of action or "ways of acting". They allow us to analyse the perceptions of the activities and responsibilities of HR managers among different organizational actors.

The empirical study examines ten companies, three of which are multinational and seven of which are Portuguese. Two of the Portuguese companies are family run. Of the ten companies, two are medium-sized companies, albeit they are leaders in their fields of business; the others are large companies (in regard to total assets, sales volume and the number of employees). The following criteria were used to select the companies: those with an HR department; those with a person responsible for HRM with the title HR manager and who effectively performed their duties; and finally, those employing an HR manager who is hierarchically and/or functionally subordinate to an administrator, general manager or superior general HR manager.

To collect the data, 257 interviews were conducted, with an average duration of 90 min. The interviewees represented different hierarchical levels, e.g. top managers, peers of HR managers and collaborators from different organizational functions, with or without managerial responsibilities. Although HR managers and their collaborators were also interviewed, given the objectives of this study, they were not considered, leaving 225 potential references. The interview guide consisted of 66 questions, among which were specific questions concerning the perception that the different organizational actors have about the ways of acting of HR managers. Only these questions were considered for the objective of this study. The first contact was made by phone with administrators and HR managers known to the first author, who was for a long period of time HR manager. The purpose of this first call was to explain

[1] This study was conducted as part of a broader project in which different themes were analysed using the same methodology. Only information corresponding to the objectives of this study was used. The broader study included the definition of research questions and the construction of the semi-structured interview guide used.

[2] According to the Oxford Dictionary, *axis* can be defined as "A straight central part in a structure to which other parts are connected" (https://en.oxforddictionaries.com/definition/axis#h70322679174160, access 04/10/2018).

Table 1 Classification of the data sample

Business clusters	Activity	Origin	Number of interviews by company	Codification of the interviewees
Multinational companies	(A) Industrial	Sweden/USA	$N = 27$	Sub. 1–27
	(B) Chemistry	Germany	$N = 27$	Sub. 28–54
	(C) Technology	Germany	$N = 28$	Sub. 55–82
National (Portuguese) companies	(D) Industrial	Portugal	$N = 25$	Sub. 83–107
	(E) Technology		$N = 25$	Sub. 108–132
	(F) Commercial		$N = 26$	Sub. 133–158
	(G) Technology		$N = 25$	Sub. 159–183
	(H) Industrial		$N = 26$	Sub. 184–209
Family companies	(I) Commercial	Portugal	$N = 24$	Sub. 210–233
	(J) Textile		$N = 24$	Sub. 234–257

the project and the main goals. I was also schedule a meeting for a comprehensive explanation of the research objectives, the data required, as well as for the definition of the number of interviews and the procedures to select the interviewees. The classification of the companies, their sectors of activity and countries of origin, the number of interviews conducted, and the codification of interviewees are described in Table 1.

The interviewees covered a broad age range, from 19 to 82 years of age, and labour seniority, ranging from 3 months to 67 years, and were at different stages of their careers. The gender distribution of the interviewees was balanced across professional categories with the exception of the administrator category, for which only three of the 24 interviewees were women. Most of the interviews were carried out between 2007 and 2012, and after transcription, given the changes in the financial and political conditions of the Portuguese context, additional interviews were conducted between 2013 and 2015, confirming the previous results.

4 Perceived Performance Axes

The perceptions analysed in this section are based on a discourse that reflects the guidelines that we designate as *axes of action* more so than a listing of HRM activities or practices. According to the words of one interviewee:

> What we have as HRM activities are those legal, administrative aspects, and the HR director is very circumscribed to these technical and administrative responsibilities… more recruitment and selection where necessary and training when there is financial capacity to do so. There could be other activities of a very different nature. Something that improved communication in the company, the ability to manage people by the chiefs, or made the company a place where it is worth returning to the next day, for example. (Subj. Manager Peer)

Table 2 Axes and scope of perceived actions

Scope	Axes of action
Essence	Reactive—Active—Proactive
Levels of action	Operational—Tactical—Strategic
Orientation	Lethargy—Accommodation—Initiative
Decision-making	Immobility—Nim—Pragmatism
Context	Local—Global
Communication	Ambiguity—Transparency
Content	Subjectivity—Objectivity
Attitude	Injustice/Partiality—Justice/Impartiality
Legal perspective	Illegalities—Pseudo-legality—Legality
Result	Inefficiency/Ineffectiveness—Efficiency/Effectiveness

This perspective reflects the existence of two types of action performed at the HRM level: existing/real actions and desirable actions, as evidenced by the perceptions expressed by the participants regarding the adoption of guidelines that influenced the performance of HR managers and HR departments (209 references). This perception is summarized as follows: "I know that the HR manager himself is aware that there is a difference between what is done and what could be done in this very important area of the company" (Subj. Technical Collaborator).

Table 2 provides a categorization of the different axes of action as perceived by the interviewees.

In labelling *axes of action*, the linguistic expressions employed by the interviewees were used to illustrate their perceptions of perceived and desirable actions by HR managers and HR departments. Therefore, the identified axes of action reveal perceptions of the roles and responsibilities of HR managers and departments and of what they could be (i.e. "the real and the desirable HRM" (Subj. Manager Peer)).

In the following subsections, the guiding categories regarding the levels of action are analysed. It is assumed that the established categories are interrelated and are not distinct but exhibit a transversality that forms channels of communication and levels of interpersonal contact, and these interactions can produce useful metaphors for the actions of HR managers and departments.

4.1 *Reactive—Active—Proactive*

These three types of performance behaviours were identified for all the studied companies. However, four trends must be noted. First, for this axis of action, the interviewees stated that the behaviours of HR managers "… *should be increasingly proactive…*" (Subj. Administrator). The proactivity of HR managers was discussed frequently by the participants (209 references of 225). However, this frequency assumes that contextual factors serve as determinants of HR managers' and departments' actions, often leading to reactive performance. The perceptions of 183 interviewees can be conveyed by the following quote: "… *the unpredictability of the object in question - person - and, on the other hand, the mutability of the contexts in which the manager of HR acts, lead to the need for an attitude that is often reactive*" (Subj. Manager Peer).

Second, the more active the actors at various levels of HRM are in terms of being good observers and listeners, organized in their actions, and able to provide and receive feedback from all internal and external interlocutors, the better the knowledge of the organization's workers in terms of hard and soft skills. In addition, when HR managers have a good understanding of the expectations that different actors have of their performance and of the business and its specificities, they are more likely to be more proactive (189 refs).

Third, although *proactivity* is considered the ideal action strategy (209 refs), the behaviour most frequently referenced as the reality is that of *reactivity* (199 refs), which is related to certain types of roles, for example "fireman" and "holy miracle worker". Reactivity largely results from a lack or even absence of an HRM strategy articulated with the company's overall strategy (176 refs) and from the fact that, contrary to rhetoric on the importance of HRM and of people, this area of management remains a second priority in organizations (168 refs). In addition, HR structures are afforded few resources and receive many requests for intervention, and there is a lack of preparedness among most Portuguese managers in HRM as well as a short-term business vision (149 refs).

Finally, there is also the perception that proactivity is linked to an anticipation of negative situations or to "black prophecies" (Subj. Technical Collaborator). These prophecies, which were mentioned by 64 of the interviewees, were always presented as non-existent in the interviewees' companies and were associated with the role of the HR manager as "the prophet of doom". For example, one interviewee stated: "*It is already known that when it is [time] to go to the HR manager, or he appears here, it is because there has been or will be disgrace. He is the true herald of doom…*" (Subj. Manager Peer).

The results of this axis of action, from the strategic HRM perspective, are in line with approaches that take a more reactive approach through which HR programmes, processes and practices contribute to the achievement of an organization's objectives (Miles and Snow 1984; Stavrou and Brewster 2005). Other proactive approaches are used by specialized HR managers during the strategic planning process, and the existing HR or their characteristics may condition an organization's strategy (Miles and Snow 1984; Uyar and Deniz 2012; Wright et al. 2001). In daily practice, the routinization of many HRM responsibilities occurs more often than not, although the expectation is that an HR department will take a proactive stance and a strategic position within its organization (Dolan et al. 2010; Uyar and Deniz 2012).

4.2 Operational—Tactical—Strategic

Actions within this axis involved the perception that only what is strategic is valued and that the aim is to view everything from a strategic level. This occurs for the sake of custom or fashion (56 refs); to ensure that one is recognized as of higher status and/or more important by other actors and organizational areas (74 refs); or because the culture and management of a company already attribute a strategic role to HRM (83 refs).

However, HRM in daily practice is still viewed as applying principles, positions and actions of a more operational nature (191 refs), in which an emphasis is placed on short-term and immediate reactive actions and interventions. Numerous organizational actors (173 refs) referred to "*… a poor understanding of the relationship between HRM's performance and the company's overall strategy*" (Subj. Manager Peer) and "*… a relative ignorance of the core of the business*" (Subj. Technical Collaborator). Such views align with those observed in previous studies (Bloom et al. 2012; Boswell 2006; Fabi et al. 2009; Martinez et al. 2013; Waldman et al. 2012).

Other interviewees argued the following:

> Of course, the ideal in HRM and in other areas, I say, for example, mine, which is even nuclear, should be more strategic… in my area that is production, as it is more objective and quantifiable, we have to be more strategic, but we still fall short. The HR manager is still very much involved in the day-to-day tasks, and as such his concern is mainly focused on the short term. (Subj. Manager Peer)

> I think HR management is often referred to as a way of increasing the credibility and the status of the area itself. However, is it necessary? People are key to the company and that is why HRM has to be strategic. (Subj. Manager Peer)

It is interesting to argue that HRM is perceived as "very operational", given that it is considered a non-objective area with few quantitative indicators (Toulson and Dewe 2004; Ulrich 1997a, b). This is viewed as a handicap by some interviewees (107 refs). According to these participants, limitations on quantification strongly limit the extent to which HRM and HR managers can act and be perceived as strategic.

Nonetheless, others viewed HRM as more strategic (121 refs) due to it being more closely related to the business, to getting to know the business better and to identifying the trajectory of the business (104 refs), as shown in previous studies (Brandl and Pohler 2010; Dolan et al. 2010). Another aspect concerns the better articulation of HRM objectives through the general objectives of the organization and the consequent operationalization of those objectives to contribute effectively to organizational outcomes (118 refs); this finding is in line with previous studies (Bloom et al. 2012; Martinez et al. 2013; Waldman et al. 2012).

The participants discussed several issues, including the following: greater proximity to people by the HR manager (97 refs); confidence in the data provided by HR managers and HR departments (94 refs); and coherence between company values and HRM practices (98 refs). According to the interviewees:

> Only by knowing the business and its evolutionary tendencies and the position that the company has at a higher level can an HR manager aspire to have strategic participation … ours does this very well (Subj. Administrator);

> In my opinion, the HRM that we have in this company is both operational and strategic… personally I do not see any problem with being both, and secondly, the mere fact that the HR structure consists of only 3 people makes it necessary to have both roles. In addition, decentralization makes it easier to concentrate on more strategic tasks, and that makes all the difference. (Subj. Manager Peer)

As highlighted in the literature (Aldrich et al. 2015), the interviewees noted that HRM must be strategic but that it is impossible not to consider the operational dimension. As highlighted by Aldrich et al. (2015, p. 109), "the degree of strategic input given to HR professionals is, in part, subject to the CEO/president's view of what HR can contribute to the realization of business plans". Nonetheless, this study demonstrates that the perspectives of different organizational actors on the strategic dimensions of HRM are paramount, as HRM affects all levels of an organization and "the choices made by the HR department on the design and implementation of 'people management' policies contribute to organizational viability and sustained competitive advantage" (Aldrich et al. 2015, p. 106).

4.3 Lethargy—Accommodation—Initiative

Although there were few references made to perceptions of *lethargy* and *accommodation* individually, the two tendencies together were mentioned by several interviewees (57 refs), as shown by the following quotes:

> There are many situations and warnings that emerge in many companies that talk about change and even the need for changes in HRM; something could even occur, but the intention is that everything stays the same…. (Subj. Administrator)

> There are companies in which the best strategy of HRM is really the accommodation, and in this company, I met an HR director who did it perfectly, he was almost invisible…. (Subj. Technical Collaborator)

Initiative was defined as a desirable behaviour and attitude (223 refs), and in many cases, it was considered a characteristic of the HR director's performance (149 refs).

> Yes, the new HR manager is a person with will and initiative; she makes things happen. (Subj. Administrator)

> Yes, I can say that even though HR is not a very numerous team, the HR manager has a huge initiative, often with very few resources, and he develops a whole set of activities that are fundamental to the different areas of our company. (Subj. Manager Peer)

However, in some cases (27 refs), *initiative* was associated with less positive aspects, as shown by the following quote:

> There is initiative, but for the benefit of oneself and one's friends … (silence), it seems that some responsible for HRM in the company divide people into three groups: *children*, *stepchildren* and *non-existent individuals*. (Subj. Technical Collaborator)

This line of action highlights the issue of initiative among HRM professionals. In this context, HR managers and HR departments can engage in other types of activities, through which they can add strategic and operational value to HRM, and can develop value-added actions for employees and the organization (Gilbert et al. 2011; Lemmergaard 2009; Schuler 1990).

Accommodating behaviour, according to the interviewees, can be interpreted as an organizational survival strategy used by an HR manager who assumes an almost non-existent presence, limiting his activity to not questioning the status quo. The professional de-individualization of the HR manager and the acceptance of a professional stance that gives voice to the *owner's voice* can create perceptions of organizational lethargy and accommodation.

4.4 Immobility—Nim—Pragmatism

In this axis of action of the HR manager, the most emphasized aspect was *pragmatism* (189 refs), which corroborates the interviewees' different views on traits valued in an HR manager. On the other hand, *nim*[3] (97 refs) is considered "*very typical of HRM and obviously nothing complimentary*" (Subj. Manager Peer). According to the interviewees, *nim* is perceived in two different ways. On the one hand, it may be tactical or infrequent "*… if it is meant to measure or consolidate a response better, but it must still be quick, which can be tolerated and accepted …*" (Subj. Manager Peer). On the other hand, when *nim* is systematic, this leads to disbelief, generates mistrust and discourages future contact or requests for help, as expressed by the following quotes:

> Nobody likes to be in the middle of a bridge without knowing which way to go (Subj. Undifferentiated Administrative Collaborator);

> Unfortunately, *nim* is not only about HR issues, but in my opinion, it is in this area of management that everything gets more complicated; the consequences can be more dire. (Subj. Undifferentiated Operational Collaborator)

Despite the critical issues that this form of action raises, the data analysis revealed that perceptions of *nim* as a mode of action can develop through different situations. An emphasis was placed on situations in which a company culture cultivates a "*chronic ambiguity*" or "*power grey zones and uncertainty*", as described by the interviewees, and it is up to the HR manager to interpret and/or agree with this mode of ambiguous management. Perceptions of *nim* as a way of acting apply to all companies, with a particular emphasis on the population of peers of HR managers (51 refs) and line managers (19 refs). This can be understood, on the one hand, by the fact that the return of HRM responsibilities is mainly experienced by these populations, making it difficult to fulfil their management obligations to their employees. On the other hand, this perception can be understood as a lack of sensitivity among these organizational actors to issues related to HRM (Bredin and Söderlund 2007; Currie and Procter 2001; Gilbert et al. 2011; Hutchinson and Purcell 2010; Perry and Kulik

[3]A person unable to say yes or no without hesitation.

2008; Renwick 2003). This lack of sensitivity can lead to expectations including the following:

> … if it is in the law, it must be black and white, and there should be no room for doubt… (Subj. Manager Peer);

> They [HR] have to be more pragmatic, more operational… We too are asked to be pragmatic in so many things where the easiest [option] was to keep postponing. (Subj. Line Manager)

The issue of immobility also emerged from the data in reference to HR managers "letting go" (13 refs). This action, in comparison to those of *nim*, was different insofar as it did not involve expectations, as reflected by the following quote: "*… we already know that we get nothing but the classic pat on the back …*" (Subj. Technical Collaborator).

4.5 Local—Global

The term "contextual" was frequently used by the participants. Of the 225 participants, all of them stressed the need for HRM to remain contextualized within the local and/or global context. The participants emphasized that HR managers should seek to frame issues based on the local contexts in which they arise and also that any implementation of actions/practices must account for the contexts considered relevant and should "anticipate the future" (Subj. Manager Peer).

Participants from multinational companies mainly focused on global performance; most of these participants were peers of HR managers (32 refs).

> The fact that we are part of a multinational company implies that HRM, and particularly the HR manager's mode of action, is global and aligned with all other HR directors of other units and with the mother company's directives. (Subj. Manager Peer of a multinational company)

However, this notion of *global* performance is always framed by the need to adapt to local contexts. Following from the previous quote:

> … although we are in tune with the mother company, the reality is that particularly in the issues of HRM, the context must always be considered: not only the context of the country and its culture but also the very particular context of the company. This, at least in our case, has been an added value because even in this area, adaptation to the context has led to situations of improvement of structural policies and even innovation in practices, which make us a case for benchmarking. (Subj. Manager Peer of a multinational company)

4.6 Ambiguity—Transparency

Concerning perceptions of actions characterized by ambiguity, the data showed different results. For companies with "bicephalic" leadership (with two HR directors) and in which HRM is centralized, more comments concerning the perceived ambiguity of the HR manager's role emerged (26 refs). In the former case (bicephalic leadership), the ambiguity stems from the different actors perceiving some overlap and a lack of definition of roles and "territorial struggles" between the two HR managers, as highlighted in the quote below. In the second case (centralized HRM), ambiguity is experienced in terms of decision-making processes, with a greater occurrence of *nim*, concomitantly with perceptions of more distance from HR managers and departments.

> The problem is that sometimes, and more than it should happen, we end up not knowing which of the HR departments we should contact. We are unsure, and even after solving a problem, it does not mean that the next time, when faced with a similar problem, we would not have the same doubts, but the company must know what it wants ... and it seems that is how it wants it. (Subj. Line Manager)

In this continuum, multinational companies have the advantage of being considerably more formalized. However, and this emerged in the three companies, when a complex or unpredictable HRM problem arises, an intense and ambiguous situation results but is usually quickly remedied, as expressed in the following quote:

> In general, this type of company deals poorly with ambiguities and seeks to plan everything, formalize everything. However, this is obviously impossible, and then I think that the fact that we are Portuguese, like MacGyver style, we managed to get everything going, we get off the ground... I think this characteristic of us in a normally planned environment but in unpredictable contexts helps us to pioneer many things in HRM. (Subj. Manager Peer)

The positive aspects emphasized included the training and the strategies that are acquired in addressing such ambiguous situations in business life (2009–2015), which is very unpredictable, ambiguous and uncertain, as mentioned in the following quote:

> The resourcefulness, combined with the ability to work, makes Portuguese emigrants out of this world, the best in the world. Obviously, being well led and having organization and planning are essential. But not everything can be planned, and not always what is planned is met, and there arises this ability of ours to go around... call it a resourcefulness, improvisation, whatever you want... we can even drive the Germans, Canadians, Norwegians to despair, but then we save the situation. Surely, you know the anecdote that says that all companies should have a Portuguese in your service ... only placed in a glass display case with a sign saying: «break in case of emergency» (laughs)... We are in the best and the worst [situation]. (Subj. Administrator) (59 refs express the same idea)

Transparency was referred to as a behavioural trait that should guide the performance of all departments, not only HRM; in fact, it "... *should be understood as a value of the company*" (Subj. Technical Collaborator; 179 refs). According to another participant, "*It would be excellent if there were transparency at all levels, but this is a dream that for some would surely turn into a nightmare ... because they would discover many bad things, many illegal schemes ...*" (Subj. Technical Collaborator).

However, it is emphasized that transparency must have limits (110 refs). These limits stem from the confidentiality associated with HRM: "*Transparency is not, nor can it be, having doors wide open. The information must be accurate, serious, useful, and transparent, but it must be disseminated in appropriate doses according to the levels of responsibility*" (Subj. Administrator).

Aligning organizational and HRM strategies and having principles, policies and practices consistent with business strategies can help increase levels of effective transparency (Bloom et al. 2012; Fabi et al. 2009; Guest 1987; Hendry and Pettigrew 1990; Martinez et al. 2013; Miles and Snow 1984; Schuler and Jackson 1987; Waldman et al. 2012).

4.7 Subjectivity—Objectivity

Regarding this line of action, the participants argued for positioning HRM and HR managers at the pole of subjectivity (193 refs), without any pejorative value attributed to them. They highlighted that subjectivity is unavoidable and results from managing people: "*… the ability to deal with the subjectivity of workers is an asset to the HR manager and department*" (Subj. Manager Peer) (148 refs). Others stated the following:

> HR managers cannot and should not stay or feel diminished by saying that this area is subjective … knowing how to deal with subjectivity is fundamental (Subj. Manager Peer);
>
> I consider HRM's subjectivity to be its main strength, and I even think that if there is something that HR mangers and departments should do, it is to place their bets on what is specific and unique to them. (Subj. Undifferentiated Operational Collaborator)

Objectivity appears to be associated with the need to assess the contributions of HRM to company outcomes and to the quantification of measurable subjects. Peers of HR managers were the group who most often referred to this quantification as ideal for HR managers and for their respective departments (91 refs). At the real/everyday level, perceptions of objective performance were considered superior in quantifiable situations (e.g. wage processing, contracting indexes, hours of training, absenteeism and turnover rates).

It is interesting to note the positions of 14 of the 24 administrators regarding objectivity. By conducting a deeper study of HRM, a better understanding of what objectivity means in HRM and a more detailed understanding of how each HR functional area can contribute to an organization's results can be obtained. Further analysis highlighted the need to operationalize performance indicators for HRM and for HR managers and departments. On this point, all 24 managers reported a need for specific metrics, also highlighting the specificities of this area and the difficulties inherent to the development of indicators, as highlighted by the following quote:

> Sometimes I feel in the management meetings a desire to demand metrics for the performance of HRM, and I feel some discomfort on the part of the HR manager who takes refuge in what he already obtains, such as the social report. In this specific case, I even consider that we have metrics that go beyond the official legal balance sheet, but I'm not even sure if it's still done. I know, however, that here in the company with this director, what is done and always has been made is an in-depth social report. Of course, at these meetings, I do not stop challenging all the managers, like "let's see, here and now, what indicators can be built"; that's where the shoe pinches, they start to move around a lot in the chairs and then… it's

> not two plus two. But this is a situation that we have to improve, and we will never know when the level of satiety is reached (laughs). (Subj. Administrator)

This perception is aligned with previous research, as the use of metrics is considered to be decisive in demonstrating HR capabilities and value added for an organization (Aldrich et al. 2015; see also Toulson and Dewe 2004; Ulrich 1997a, b, c).

4.8 Injustice/Partiality—Justice/Impartiality

Although all interviewees noted that it is ideal to promote justice in addressing HRM issues (225 refs), 86 participants made direct references to situations of injustice perpetrated by those who have to manage or lead people. It was noted that:

> In this company, many things are done well and many things bad … in this aspect, we should be like most. Now, with regard to the management of people, there are double standards, which is not justified, or better only justified if in fact one person is excellent and another is intractable. In this case, there must be differentiation, but sometimes it is the opposite; the good employee is the one who is discriminated against… Boot-licking is always gotten away with, and why? Well, that too I would like to understand! (Subj. Undifferentiated Operational Collaborator);

> Leaving a worker, for years, facing a wall with no assigned function, with nothing to do, is what? Where is HR? What does the HR manager do? The HR manager has conversations to get the person to accept ridiculous compensation and leave. Of course the person I am talking about is already numb and can handle everything … now she got tired and put the company in court … Sometimes a mouse and a few ants shake a group of elephants. A shame…. (Subj. Technical Collaborator)

The issue of partiality and impartiality was discussed similarly by the interviewees, as shown by the following quote:

> …Good professional performance must be distinguished from poor professional performance. Professional pride, the effective commitment of people must be distinguished and rewarded; we cannot treat the same what is different. (Subj. Manger Peer)

The interviewees referred to impartiality as a form of justice, as shown by the following quote (corroborated by 112 refs):

> … I do not agree with "equal work, equal pay" because the work may even be the same, but the performances are different, it is undeniable. If there is a colleague who does the same as me and is better than me, then he should receive more, receive a prize and I should not! We are talking about justice and injustice, and there are still many injustices. (Subj. Undifferentiated Operational Collaborator)

4.9 Illegalities—Pseudo-legality—Legality

Compliance with legality emerged as a form of action that was desired by the participants, many of whom consider this to be the form of action adopted by HRM hierarchies, mainly as an example of ethical values. Pseudo-legality was considered a form of action that was largely not assumed, mostly by administrators and managers' peers. The references regarding this form of dubious action involved issues of labour law and arose primarily at the contracting level, more specifically, as there are some legal rules that must be fulfilled in that process. As mentioned:

> In my area, I have to dig up information and invent to get results. My fellow HR manager has to do the same. I talk to friends of mine from other companies, and I think we are very legalistic; if the HR department only sees illegalities and fines, there will be ways to circumvent the law .... (Subj. Manager Peer)

The interviewees from multinational companies considered the issue of legality to be one of the most significant problems faced in Portugal. Curiously, the three administrators already knew of the expression *chico-esperto* [clever clogs]: "*They say it's a country of chico-espertos (laughs), in my country there are also ... (laughs), sorry, no joke*" (Subj. Administrator). They also referred to the idea that "*... working within the law is difficult, but if it were easy, we would certainly not be needed (laughs) ... but working without law is impossible*" (Subj. Administrator).

Other participants from three national companies (27 refs) addressed this issue using ethics-based terminology that reflected the continuum of *unethical behaviours/ethically irreprehensible behaviours*, as shown by the following quote:

> I'm even afraid to speak in ethics, in a world where the word of a man is worthless or worth very little, which is basically the same thing. If there is one thing I like to do, it is watch people, and I try hard to understand how the competition works. I hear and see things that are absolutely unfortunate and unthinkable from an ethical point of view. What amazes me is seeing certain companies referred to as the best companies to work for! Someone is deceived or absolutely mistaken. I do not understand! (Subj. CEO)

Legality is relevant to the functioning of HRM, HR managers and HR departments, according to our interviewees and to the literature (Yamamoto 2009). Authors such as Coda et al. (2009) and Yamamoto (2013) emphasize how HRM boards are institutionalized, formalized, and built and how expectations surrounding their performance are developed, where ethical standards and issues of legality are presented as principles to follow.

4.10 Inefficiency/Ineffectiveness—Efficiency/Effectiveness

Although there was a clear appreciation of efficient and effective actions (216 refs) in HRM and in other organizational areas, the effectiveness and directed behaviour of HR managers in the *practical* resolution of situations were perceived as more relevant by the interviewees.

The main barriers to effective action described are reflected by the following quotes: "*there are still many nim*" (Subj. Line Manager) and "*Problems are postponed, and things do not go as they should; we postpone decisions and we lose opportunities*" (Subj. Manager Peer). The very nature of HRM is considered to be complex: "*when you ask for a simplification of processes, sometimes we tend to connect the complicometer [ability to complicate], and this removes the discernment to do what has to be done. It reduces efficacy and generates dysfunctions and entropies in the right way of doing things, and in that sense, efficiency is lost*" (Subj. Administrator). In addition, several interviewees emphasized that, in an exaggerated way, HRM involves "… *effectiveness and efficiency to the limit*" (Subj. Technical Collaborator). Interestingly, one administrator referred to these excesses and inadequacies as follows:

> Everything that is exaggerated is harmful. Using medical language, one can say, as Paracelsus said, that "the dose is the poison". The role of our HR manager is to diagnose and act… to detect the inefficiencies in his area and from there to adopt the correct management measures. This is how the HR manager has positioned himself, and he has done very well. Sometimes we have to sacrifice efficiency for efficacy, and, as the Spaniards say, «no pasa nada» [nothing happens]. If I have a fire that I can control with a fire extinguisher and if I know how to use it, I do not set up a meeting to discuss how to extinguish the fire …. (Subj. Administrator)

5 Discussion

Through the perceptions of different organizational actors, this study aimed to better understand and characterize HR managers' actions at the strategic, tactical and operational levels. The existing perceptions of HR managers are related to perceptions of the direction of HR, especially at the lowest levels of the organizations analysed, and are influenced by the perceived axes of action adopted by all actors involved in HRM. This finding is confirmed by the gap identified in the discourse of several of our interviewees between the current and the desired HRM practices. This gap has been widely reported in HRM research (Cabral-Cardoso 2004; Dolan et al. 2010; Guest 1987, 2010; Mirfakhar et al. 2018; Pfeffer 1995). More interesting, in this study, is the fact that this gap is usually identified in reference to the HRM practices of other companies and not in reference to the interviewee's company. In addition, the axes of action perceived by the participants are somewhat similar, which can be explained by the transversality attributed to HRM in terms of principles, policies and practices. Although classified under different scopes, such axes of action are not independent and autonomous; there are common features between them.

The study shows that there are eight axes of action, each one representing a continuum, with our interviewees highlighting their views of what the real and desirable practices are. In terms of effective axes of action, the results highlight that HR managers are reactive in essence. Their level of action is mainly operational but also tactical. HR managers are perceived as adopting an accommodated orientation, in an attempt to maintain the status quo and professional survival, although some initiative is recognized. In terms of decision-making, HR managers are characterized by immobility and pragmatism, but above all, they are viewed as incapable of making firm decisions (*nim*). Within a company and its context, HR managers are perceived as capable of considering both local and global contexts. In terms of communication,

HR managers are viewed as ambiguous, and concerning content, they are viewed as subjective. Regarding their attitudes, they are perceived as acting with some level of injustice and partiality combined with a position of pseudo-legality. Finally, in terms of results, HR managers are viewed to a certain degree as inefficient and ineffective. These results are aligned with the development of HRM in Portugal, reflecting the move from a purely bureaucratic and administrative role to a people-centred approach to management (Cabral-Cardoso 2004).

The function of HR manager is a reality mainly in Portuguese medium and large companies, as in our sample, and interviewees are very clear about what they consider to be desirable axes of HR manager actions, and these behaviours would support more positive perceptions of HR managers and departments. These desirable axes are as follows: proactive; strategic and operational; demonstrating initiative; pragmatic; simultaneously global and local (*glocal*); transparent; objective; just and impartial; legal; and efficient and effective. These results are consistent with the GLOBE analysis of Portugal, where future orientation is valued more than the actual practice and where power distance is a reality both in Portuguese society and organizations (House et al. 2004).

Three major axes of action emerge from this analysis as per Fig. 1: principles of practical action, principles of communication and ethical values. The principles of practical actions are related to the most relevant aspects of everyday life of the different stakeholders of HRM (Buyens and De Vos 2001; Dolan et al. 2010; Guest 2010; Tsui 1990; Ulrich 1997c, 1998a; Waldman et al. 2012). That is, better systems of recruitment and selection, reception and integration, evaluation and performance management, career management, remuneration policies, benefits and incentives, coordination and motivation of people and teams, and effectiveness and efficiency in results. However, the different stakeholders can only effectively perceive this if the HR managers and department adopt a transparent and objective communication system. Importantly, and although not particularly highlighted by the literature (Bloom et al. 2012; Dolan et al. 2010), ethical values are perceived as very important in this study and need to be connected with the practical actions and communication principles of HR managers. It is of significant importance that the different stakeholders perceive HRM as fair, coherent, impartial and capable of fostering the happiness and optimism of its workers.

6 Conclusion

The results of this study demonstrate that when building their identity and credibility within organizations, HR managers must consider the perceptions of different organizational actors. The ways in which HR managers are perceived by different organizational actors pose a challenge and affect the roles and influence of HR managers and departments at the strategic level of organizations. On one hand, the organization aims to have collaborators, from different levels, committed and motivated with their work and with the organization, and flexible to adapt when confronted with the

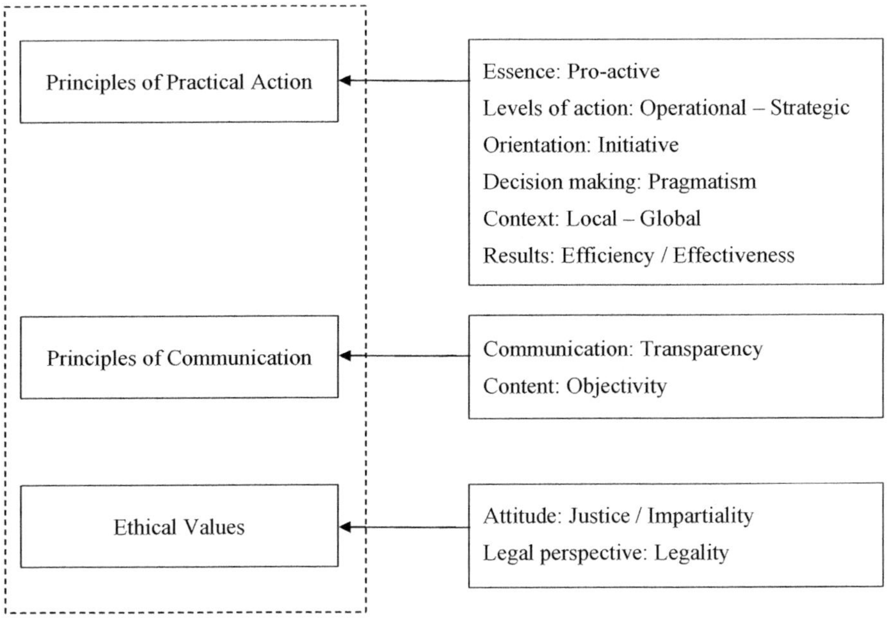

Fig. 1 Framework on axes of action of HR managers—desirable actions

challenges inherent to processes of change (Hendry and Pettigrew 1990; Guest et al. 2003; Cooke and Saini 2010). On the other, the different organizational actors aim to have their personal and professional goals satisfied by the organization to which they belong (Bloom et al. 2012; Brandl and Pohler 2010; Dolan et al. 2010; Guest 2010; Tsui and Milkovich 1987; Ulrich 1997c, 1998a; Waldman et al. 2012).

The study has demonstrated that three major axes of action emerge from the perception of the different stakeholders of HRM: principles of practical action, principles of communication and ethical values. It is paramount that HR managers are aware of these perceptions and adopt the principles to help enhance the role HRM assumes within an organization. This will help them in proactively anticipate and solve problems while adding value and creating competitive and distinctive organizational advantages in the short, medium and long terms (Aldrich et al. 2015; Bloom et al. 2012; Cabral-Cardoso 2004; Farndale 2005; Paauwe and Boselie 2003; Waldman et al. 2012; Yamamoto 2013). The identified axes of action, in this study, may contribute to overcome some of the obstacles to the development of HRM. One of those obstacles is created by the ambiguities and inconsistencies at various levels of HRM and notably in terms of the nature and exercising of the authority of HR managers (Bloom et al. 2012; Cabral-Cardoso 2004; Davila et al. 2010; Mirfakhar et al. 2018; Torrington 1998). Of significant relevance is the capacity of HR managers to keep a positive relationship and high-quality communication with HR stakeholders as well as high ethical values (Klaas et al. 2012; Mirfakhar et al. 2018). In the specific Portuguese context, a focus of HR managers on the three main axes of action may help them to be listened in the organization, to set the HRM agenda, and improve their credibility in organizations (Cabral-Cardoso 2004).

The study reveals a concern with the activity of HRM and the HR manager, which can be genuine, instrumental, strategic or operational but always related to the specific interests (personal and/or professional) of the different organizational actors. This finding is aligned with previous studies, which highlight the relevance of HRM to keep a positive relationship and high-quality communication with HR stakeholders and consider their specific needs (Klaas et al. 2012; Mirfakhar et al. 2018). Therefore, several practical implications for HR managers in their activities within organizations emerge. These implications are described below and are related to the need for HR managers and departments to develop approaches and behaviours that promote organizational empathy for all actors of an organization. First, it is necessary to understand and characterize the perceptions of different organizational actors. Second, it is paramount to consider the personal and professional expectations of organizational actors and to incorporate them into HRM policies. The objective is for HRM practices to more closely reflect the practices desired by the different actors without detracting from its role in organizations and while making its guidelines clear. Third, the organizational decision-making process should be contextualized by creating conditions that promote the involvement of different organizational actors whenever possible. Fourth, training and communication systems that favour the development of proactive, strategic and systemic ways of acting in an organization must be developed. Fifth, the hard and soft skills in an organization, as well as the main gaps, should be identified while taking into account the vision, mission and strategic objectives of the organization. Sixth, internal marketing programmes for HRM activities involving different organizational actors must be developed, so that such actors can also contribute to improve HRM. This can also help reveal facilitators and obstacles that the implementation of HRM policies and practices may face when encountered by the different organizational actors in real life. All these implications for HR managers are related to the fact that HR stakeholders' expectations are derived from their role or functional responsibilities, their positions in the organizational hierarchy, but also from their career and personal circumstances, as argued by Tsui and Milkovich (1987).

This study contributes to the HRM literature by exploring an important yet underdeveloped field of research on how HR managers are perceived within organizations, and specifically within the Portuguese context (Aldrich et al. 2015; Brandl and Pohler 2010; Farndale 2005; Markoulli et al. 2017; Uyar and Deniz 2012; Yamamoto 2013). By focusing on the perceptions of different organizational actors, this study contributes to the debate on how the perceptions of organizational members affect the roles and impact that HR professionals can have at the strategic level of organizations. This contribution is of particular relevance, and is aligned with previous research, which argued that a major source of complexity for HR managers and departments is the plurality of stakeholders (Klaas et al. 2012; Mirfakhar et al. 2018; Tsui 1990; Tsui and Milkovich 1987). For HR practitioners, this study contributes to a better understanding on how stakeholders' expectations define who and how HR must satisfy to help the firm succeed (Ulrich et al. 2017), since HR managers and departments serve multiple stakeholders or constituencies, with multiple and diverse objectives (Tsui 1990).

Notwithstanding the contribution of this study, this is focused on the Portuguese context and there is still the need to better understand the perceptions of HR managers' activities, which is a topic that has been of genuine interest to HRM scholars for many years now and still needs further research, in different macro-, mezzo- and micro-organizational contexts.

Acknowledgements The second author acknowledges that this study was conducted at Research Center in Political Science (UID/CPO/00758/2013), University of Minho, and supported by the Portuguese Foundation for Science and Technology and the Portuguese Ministry of Education and Science through national funds.

References

Aldrich, P., Dietz, G., Clark, T., et al. (2015). Establishing HR professionals' influence and credibility: Lessons from the capital markets and investment banking sector. *Human Resource Management, 54*(1), 105–130.

Bloom, N., Genakos, C., Sadun, R., et al. (2012). Management practices across firms and countries. *Academy of Management Perspectives, 26*(1), 12–33.

Boon, C., Den Hartog, D. N., Boselie, P., et al. (2011). The relationship between perceptions of HR practices and employee outcomes: Examining the role of person–organisation and person–job fit. *The International Journal of Human Resource Management, 22*(1), 138–162.

Boselie, P., & Paauwe, J. (2005). Human resource function competencies in European companies. *Personnel Review, 34*(5), 550–566.

Boswell, W. (2006). Aligning employees with the organization's strategic objectives: Out of 'line of sight', out of mind. *The International Journal of Human Resource Management, 17*(9), 1489–1511.

Brandl, J., & Pohler, D. (2010). The human resource department's role and conditions that affect its development: Explanations from Austrian CEOs. *Human Resource Management, 49*(6), 1025–1046.

Bredin, K., & Söderlund, J. (2007). Reconceptualising line management in project-based organisations: The case of competence coaches at Tetra Pak. *Personnel Review, 36*(5), 815–833.

Brewster, C. (1997). A profissão de gerir pessoas e talentos. In *18° Congresso Europeu de Recursos Humanos* (pp. 7–13). Lisboa: APG.

Buyens, D., & De Vos, A. (2001). Perceptions of the value of the HR function. *Human Resource Management Journal, 11*(3), 70–89.

Cabral-Cardoso, C. (2004). The evolving Portuguese model of HRM. *The International Journal of Human Resource Management, 15*(6), 959–977.

Coda, R., César, A., Bido, D., et al. (2009). Strategic HR? A study of the perceived role of HRM departments in Brazil and Peru. *BAR—Brazilian Administration Review, 6*(1), 15–33.

Cooke, F. L., & Saini, D. S. (2010). (How) does the HR strategy support an innovation oriented business strategy? An investigation of institutional context and organizational practices in Indian firms. *Human Resource Management, 49*(3), 377–400.

Corbin, J., & Strauss, A. (2008). *Basics of qualitative research: Techniques and procedures for developing grounded theory*. Thousand Oaks, CA: Sage.

Currie, G., & Procter, S. (2001). Exploring the relationship between HR and middle managers. *Human Resource Management Journal, 11*(3), 53–69.

Davila, A., Foster, G., & Jia, N. (2010). Building sustainable high-growth startup companies: Management systems as an accelerator. *California Management Review, 52*(3), 79–105.

Dewettinck, K., & Vroonen, W. (2017). Antecedents and consequences of performance management enactment by front-line managers. Evidence from Belgium. *The International Journal of Human Resource Management, 28*(17), 2473–2502.

Dolan, S. I., Eisler, R., & Raich, M. (2010). Managing people and human resources in the XXI Century. Shifting paradigms, emerging roles, threats and opportunities. *Effective Executive, 13*(6), 39–46.

Fabi, B., Raymond, L., & Lacoursière, R. (2009). Strategic alignment of HRM practices in manufacturing SMEs: A gestalts perspective. *Journal of Small Business and Enterprise Development, 16*(1), 7–25.

Farndale, E. (2005). HR department professionalism: A comparison between the UK and other European countries. *The International Journal of Human Resource Management, 16*(5), 660–675.

Gannon, J. M., Roper, A., & Doherty, L. (2015). Strategic human resource management: Insights from the international hotel industry. *International Journal of Hospitality Management, 47,* 65–75.

Gilbert, C., De Winne, S., & Sels, L. (2011). The influence of line managers and HR department on employees' affective commitment. *The International Journal of Human Resource Management, 22*(8), 1618–1637.

Guest, D. (1987). Human resource management and industrial relations. *Journal of Management Studies, 24*(5), 503–521.

Guest, D. (2010). Human resource management and performance: Still searching for some answers. *Human Resource Management Journal, 21*(1), 3–13.

Guest, D., Michie, J., Conway, N., & Sheehan, M. (2003). Human resource management and corporate performance in the UK. *British Journal of Industrial Relations, 41*(2), 291–314.

Hendry, C., & Pettigrew, A. (1990). Human resource management: An agenda for the 1990s. *The International Journal of Human Resource Management, 1*(1), 17–43.

Hennessy, J., & McCartney, C. (2008). The value of HR in times of change. *Strategic HR Review, 7*(6), 16–22.

House, R. J., Hanges, P. J., Javidan, M., et al. (2004). *Culture, leadership, and organizations: The GLOBE study of 62 societies*. London: Sage.

Hutchinson, S., & Purcell, J. (2010). Managing ward managers for roles in HRM in the NHS: Overworked and under-resourced. *Human Resource Management Journal, 20*(4), 357–374.

Ibănescu, A. (2015). How do novice specialists in human resources develop their professional identity? *Social and Behavioral Sciences, 187,* 596–600.

Klaas, B. S., Semadeni, M., Klimchak, M., & Ward, A.-K. (2012). High-performance work system implementation in small and medium enterprises: A knowledge-creation perspective. *Human Resource Management, 51*(4), 487–510.

Laperrière, A. (2010). A teorização enraizada (grounded theory): procedimento analítico e comparação com outras abordagens similares. In J. Poupart, J. P. Deslauriers, L. H. Groulx, et al. (Eds.), *A Pesquisa Qualitativa: Enfoques Epistemológicos e Metodológicos. Tradução de Ana Cristina Nasser* (pp. 353–385). Petrópolis, RJ: Vozes.

Lemmergaard, J. (2009). From administrative expert to strategic partner. *Employee Relations, 31*(2), 182–196.

Lo, K., Macky, K., & Pio, E. (2015). The HR competency requirements for strategic and functional HR practitioners. *The International Journal of Human Resource Management, 26*(18), 2308–2328.

Markoulli, M. P., Lee, C., Byington, E., et al. (2017). Mapping human resource management: Reviewing the field and charting future directions. *Human Resource Management Review, 27*(3), 367–396.

Martinez, L. F., Ferreira, A. I., & Lopes, M. P. (2013). *Gerir Pessoas. 14 Exemplos de Boas Práticas em Portugal*. Lisboa: Edições Sílabo.

Meijerink, J. G., Bondarouk, T., & Lepak, D. (2016). Employees as active consumers of HRM: Linking employees' HRM competences with their perceptions of HRM service value. *Human Resource Management, 55*(2), 219–240.
Miles, R. E., & Snow, C. C. (1984). Designing strategic human resources systems. *Organizational Dynamics, 13*(1), 36–52.
Mirfakhar, A. S., Trullen, J., & Valverde, M. (2018). Easier said than done: A review of antecedents influencing effective HR implementation. *The International Journal of Human Resource Management*. Available online. https://doi.org/10.1080/09585192.2018.1443960.
Myers, M. D. (2011). *Qualitative research in business and management*. London: Sage.
Ng, L. C. (2011). Best management practices. *Journal of Management Development, 30*(1), 93–105.
Paauwe, J., & Boselie, P. (2003). Challenging 'strategic HRM' and the relevance of the institutional setting. *Human Resource Management Journal, 13*(3), 56–70.
Pernkopf-Konhäusner, K., & Brandl, J. (2011). Variations in evaluative repertoires: Comparing employee perspectives on training and development in Germany and Russia. *Personnel Review, 4*(5), 589–606.
Perry, E. L., & Kulik, C. T. (2008). The devolution of HR to the line: Implications for perceptions of people management effectiveness. *The International Journal of Human Resource Management, 19*(2), 262–273.
Pfeffer, J. (1995). Producing sustainable competitive advantage through the effective management of people. *Academy of Management Perspectives, 9*(1), 55–69.
Pfeffer, J. (1998). *The human equation*. Cambridge, MA: Harvard Business School Press.
Piening, E. P., Baluch, A. M., & Ridder, H.-G. (2014). Mind the intended-implemented gap: Understanding employees' perceptions of HRM. *Human Resource Management, 53*(4), 545–567.
Renwick, D. (2003). Line manager involvement in HRM: An inside view. *Employee Relations, 25*(3), 262–280.
Rivero, A. G., & Dabos, G. E. (2017). Gestión diferencial de recursos humanos: una revisión e integración de la literatura. *Estudios Gerenciales, 33*(142), 39–51.
Schuler, R. S. (1990). Repositioning the human resource function: Transformation or demise? *Academy of Management Perspectives, 4*(3), 49–60.
Schuler, R. S., & Jackson, S. E. (1987). Linking competitive strategies with human resource management practices. *Academy of Management Perspectives, 1*(3), 207–219.
Scullion, H., & Starkey, K. (2000). In search of the changing role of the corporate human resource function in the international firm. *The International Journal of Human Resource Management, 11*(6), 1061–1081.
Sheehan, C., De Cieri, H., Cooper, B., et al. (2014). Exploring the power dimensions of the human resource function. *Human Resource Management Journal, 24*(2), 193–210.
Stainback, K., Tomaskovic-Devey, D., & Skaggs, S. (2010). Organizational approaches to inequality: Inertia, relative power, and environments. *Annual Review of Sociology, 36*(1), 225–247.
Stavrou, E. T., & Brewster, C. (2005). The configurational approach to linking strategic human resource management bundles with business performance: Myth or reality? *Management Revue, 16*(2), 186–201.
Stone, D. L., & Deadrick, D. L. (2015). Challenges and opportunities affecting the future of human resource management. *Human Resource Management Review, 25*(2), 139–145.
Strauss, A., & Corbin, J. (1998). *Basics of qualitative research*. London: Sage.
Torrington, D. (1998). Crisis and opportunity in human resources management. The challenge for the personnel function. In: P. Sparrow & M. Marchington (Eds.), *Human resource management. The new agenda* (pp. 23–36). New York: Pitman Publishing.
Toulson, P. K., & Dewe, P. (2004). HR accounting as a measurement tool. *Human Resource Management Journal, 14*(2), 75–90.
Truss, C., Gratton, L., Hope-Hailey, V., et al. (2002). Paying the piper: Choice and constraint in changing HR functional roles. *Human Resource Management Journal, 12*(2), 39–63.
Tsui, A. (1990). A multiple-constituency model of effectiveness: An empirical examination at the human resource subunit level. *Administrative Science Quarterly, 35*(3), 458–483.

Tsui, A., & Milkovich, G. (1987). Personnel department activities: Constituency perspective and preferences. *Personnel Psychology, 40*(3), 519–537.
Tung, R. L. (2016). New perspectives on human resource management in a global context. *Journal of World Business, 51*(1), 142–152.
Ulrich, D. (1997a). Judge me more by my future than by my past. *Human Resource Management, 36*(1), 5–8.
Ulrich, D. (1997b). HR of the future: Conclusions and observations. *Human Resource Management, 36*(1), 175–179.
Ulrich, D. (1997c). *Human resource champions: The next agenda for adding value and delivering results*. Boston: Harvard Business School Press.
Ulrich, D. (1998a). A new mandate for human resources. *Harvard Business Review, 76*(1), 124–134.
Ulrich, D. (1998b). The future calls for change. *Workforce, 77*(1), 87–91.
Ulrich, D., & Dulebohn, J. H. (2015). Are we there yet? What's next for HR? *Human Resource Management Review, 25*(2), 188–204.
Ulrich, D., Kryscynski, D., Brockbank, W., & Ulrich, M. (2017). *Victory through organization: Why the war for talent is failing your company and what you can do about it*. New York: McGraw-Hill.
Urquhart, C. (2013). *Grounded theory for qualitative research: A practical guide*. London: Sage.
Uyar, A. S., & Deniz, N. (2012). The perceptions of entrepreneurs on the strategic role of human resource management. *Procedia—Social and Behavioral Sciences, 58*, 914–923.
Waldman, D. A., De Luque, M. S., & Wang, D. (2012). What can we really learn about management practices across firms and countries? *Academy of Management Perspectives, 26*(1), 34–40.
Wright, P. M., Dunford, B. B., & Snell, S. A. (2001). Human resources and the resource based view of the firm. *Journal of Management, 27*(6), 701–721.
Yamamoto, H. (2009). *Retention management of talent: A study on retention in organizations*. Tokyo: Chuokeizai-Sha.
Yamamoto, H. (2013). The relationship between employees' perceptions of human resource management and their retention: From the viewpoint of attitudes toward job-specialties. *The International Journal of Human Resource Management, 24*(4), 747–767.

About Competencies, Creativity, and Innovation in the Portuguese Textile and Clothing Sector

Carolina Feliciana Machado and Rosa Maria Maia Miranda

Abstract According to the 2020 Strategy, innovation and human resources (HR) qualification are critical in organizations' competitiveness. Portuguese textile and clothing sector(TCS) currently occupies, respectively, the eighth and fifth place in the textile and clothing industry. Considered a "traditional" sector, in the last few years, has faced difficulties as well as opportunities, through products and processes innovation. Thanks to their skills, know-how, creativity, and abilities HR are crucial for organization's innovation. We propose to research HR' skills, qualifications, and creativity within the TCS, measuring to what extent HR qualification and creativity play a role in the organization's innovative ability. Aiming the success of 2020 Strategy, the bet on HR and innovation is a relevant topic, to be considered by TCS organizations. A sample was selected of companies within the TCS with innovative abilities, which, by their practices, may set an example for other organizations, reaching for high levels of innovation and HR management.

Keywords Competencies · Creativity · Innovation · Human resource management · Portugal · Textile and clothing sector

1 Introduction

In nowadays economic and political context, more than never the organization competitiveness is vital in order to give to the client a product/service in time, assuring its profitability, regardless the organization position in the supply chain. An organization competitiveness is its ability to go further, being more attentive to the market needs, optimizing resources and earning the market position. One of the factors that lead to competitiveness is the organization ability to be different from its competitors and hold their own, being innovation a critical option in current days, in a context of competitiveness and crisis in the different world economies. This way, besides other variables, qualified human resources (HR) and resources management

C. F. Machado (✉) · R. M. M. Miranda
School of Economics and Management, University of Minho, Braga, Portugal
e-mail: carolina@eeg.uminho.pt

C. Machado and J. P. Davim (eds.), *Entrepreneurship and Organizational Innovation*, Management and Industrial Engineering, https://doi.org/10.1007/978-3-030-19289-1_6

policies, along with innovation, promote a competitive advantage in relation to competitors, allowing to get a market share and reach their aims (Noe 2010). Innovation, in an organization, appears as a possible way to differentiate and assert itself in a so demanding global market. Being HR one of the most vital resources to drive innovation and differentiation, they need to be managed, trained, and developed in order to positively contribute to the organization. On the other hand, will it be enough to have trained and developed HR to promote innovation? Does innovation occur only with creative HR?

We are facing a period of instability and economic crisis with European economies becoming more vulnerable to external factors, such as oil, currency appreciation, financial system, reflecting difficulties to affirm itself, while economies of emerging countries are growing.

Through innovation, organizations can compete in more exigent markets with add value products and services, having the best HR in the organization key roles.

On the other hand, Portugal positioning in the research, development, and innovation field, established in the 2020 Strategy, is still at 50% of the agreed targets.

By comparison, countries like Japan or South Korea invest about twice than Europe in what concerns costs in R&D relative to GDP, which places us, globally, in a position less innovative than other countries, although with a significant growth (European Commission 2011).

Additionally, European Union remains competitive worldwide as it has the greatest number of researchers in the world, at the same time that in what concerns scientific production and research and technologies patenting is the second biggest research center, just following the USA. However, in some areas, is still behind some world competitors and its competitive position in relation to other countries, has worsened, existing a performance gap in many research areas (European Commission 2011).

Besides consider research, development, and innovation themes, the 2020 Strategy also considers HR qualifications, as another crucial area, to reinforce the European Union action and positioning. Therefore, it is a priority to organizations to invest in innovation and in HR qualifications.

Textile and clothing sector (TCS) is a sector with a high level of representation to the Portuguese economy. In 2013 and according to *Associação Têxtil e Vestuário de Portugal*—ATP (Portuguese Textile and Clothes Association) (2013), TCS still has a crucial role in the national economy, representing 9% of the total exports, 20% of employment in the manufacturing industry, 8% of the businesses volume in the manufacturing industry, and 8% of production in this industry.

TCS organization's cluster gives a particular focus in innovation and differentiation, presenting innovative products in recognized international exhibitions that occur in different European cities such as *Heimtêxtil* in Frankfurt, *ISPO* in Munique, *Texworld* and *Maison Object* in Paris, and many other expertise exhibitions. However, being a traditional and labor-intensive sector, the existence of qualified HR is crucial, aiming, with the present research, present to the sector results that will allow to make it easier the implementation of HR practices in the sector which lead with success to the established challenges of the Horizon 2020 Strategy about HR innovation and qualification.

The main aim of this paper is, this way, to contribute to the business and academic community, studying the relation of the innovative capacity of an organization with the HR competencies as well as in what way HR qualifications and creativity can influence the organization capacity to innovate. As intermediate aims we can consider, among others, to research the relation between HR creativity and qualifications (More creative persons have more qualifications?), and to research what are the used tools in creativity as a tool to obtain innovation (Will be the innovation result of chance or result of systematic practices and creativity stimulus of qualified HR?). The key issues that we are going to research are HR competencies and qualifications, creativity, and innovation, in a sample of organizations from the TCS, considered innovative and with innovation practices that accepted to collaborate in this research.

2 Literature Review

2.1 Competitiveness

Be competitive in the global market is more than an aim; it is an ambition of any continent, country, region, or organization. Globalization phenomenon is located in our society and in our economies, since acquiring a product or a service, at any time and in any place in the world, is a reality to which organizations and costumers are already familiar. In this new reality, allied to this critical and severe economic scenario that we are facing, it is essential to be competitive, both to a nation or an organization, which provides a high competitiveness, resulting that only competitive organizations will have a place in an economy increasingly complex, demanding, unstable, and dynamic.

But, what competitiveness means? How is it measure? Any ordinary citizen will say that it is easy to give a definition of competitiveness. However, this definition is not objective, not existing a universal definition, with different authors arguing that not always is easy to define the competitiveness of a nation or an organization.

According to Farole et al. (2010), competitiveness is a vague concept, open to different interpretations. Solleiro and Castanón (2005) refer that competitiveness is a complex concept that has been studied with different approaches and subject areas, not being possible, however, to accept one definition.

Buckley et al. (1988), by their way, considered that competitiveness could refer to an organization or a nation competitiveness, setting different measure categories to competitiveness, in what concerns the organization or the nation competitiveness.

In a relatively more recent definition, Artto (referred by Ulengin et al. 2011) considers nation competitiveness as the degree in which a nation can produce, under correct and free market conditions, products and services internationally, while at the same time expand real benefits to citizens. This definition leads us to different perspectives of analysis, such as in the area of corporate social responsibility (*to produce under correct and free conditions*), or in the area of consumption and costumers, who receive from organizations useful products to facilitate the day-to-day life.

According to OECD (2010), competitiveness is the measure of advantage or disadvantage of a country which sells its products in international markets, in other words, a country competitiveness is so great how higher is its capacity to sell products and services in global markets, to different clients and costumers.

To the World Economic Forum (2014), competitiveness is defined as the group of institutions, policies, and factors that determinate the productivity level of a country.

Reinforcing their position, Buckley et al. (1988) considered that organizations should present their products and services in diversified and with add value markets. Therefore, presentation of new products happens, among other factors, when existing investment in research, development, and innovation, as well as HR trained and involved in the organization (2010).

An organization acquires sustainable competitive advantages when has something that sets it apart from its competitors, present or potential, implementing and creating valuable strategies to the different stakeholders at the same time that their competitors do not have the necessary ability to do it in a continuous way (Barney 1991). Being this position still valid nowadays, Dessler (2013) highlights that the best competitive advantage of an organization presently is, frequently, the human capital. In other words, employees with the necessary knowledge, qualifications, committed with the organization and their jobs, work hard. This way, organizations do the difference by the human capital they have and by its capacity to make these resources profitable.

2.2 Human Resources and Its Link with Competitiveness

Being resources one of the key supports that allow to evaluate economies in what concerns its competitiveness, some issues arise, namely what are resources? What is its classification? What is its relevance in organizations?

About an organization's resources, we can say that they include all necessary abilities to the business functioning, development, and improvement, namely, physical/tangible resources, such as buildings, machines, technologies and information systems; as well as intangible resources, like knowledge, technics and production methods, technical documents, etc.

According to Barney (1991), not all resources have the same relevance and contribute equally to the sustainable competitive advantage of an organization, establishing, this way, four attributes that resources need to fulfill in order to be considered strategic in an organization:

- value to the organization—does the resource generate added value, enable gains, exploit new opportunities, and minimize threats?

- rarity—is there a small number of resources or is there an organization with this resource?
- immeasurable—the value-added resource is rare, can it be reproduced and/or imitated? and
- organizational—is the organization equipped and organized to maximize its resources?

Thus, as argued by Barney (1991), Solleiro and Castanón (2005), Maier et al. (2014), human capital is essential to the development of any business, depending on its success and competitiveness of this intellectual capital. Thus, it is fundamental to manage human capital, in order to promote the application of its competencies, to the benefit of the organization.

The relationship between human resource management (HRM) and organizational performance has long been part of academic research, having several authors found similar conclusions. Indeed, there is a strong relationship between HRM and organizational performance in several areas: organizational performance (Becker and Gerhart 1996), productivity and financial performance (Huselid 1995), HR capacities and competitive advantage (Auw 2009), organization strategy, and HR and performance results (Buller and McEvoy 2012). The performance of organizations is related to the ability to do different, and by much technology that organizations have, among other resources, HR are different from organization to organization, as well as their ability to innovate.

According to the Innovation Union Report (European Commission 2013), in highly innovative economies, it is necessary to invest and "build" qualified and motivated HR which combining specific and general knowledge must be able to contribute to innovation. Being the human capital essential to the functioning of an organization, it is necessary to manage it today, to guarantee a competitive company tomorrow.

Due to its critical relevance, what do we understand by human resources management? What is its importance in organizations? What is the meaning and how to acquire competencies? How to develop HR and their competencies?

2.3 Human Resource Management and Competencies

According to Dessler (2013), HRM is the process of acquiring, training, evaluating, and compensating employees, paying attention to labor relations, safety and health, and concepts of justice. This same author (2013) considers that strategic HRM means formulating and executing HR practices and policies that produce employee competencies and behaviors that these companies need in order to achieve their strategic objectives. In other words, it is to assume that HR must be at the level of any other strategic area of an organization and must be prepared, qualified, and competent to follow and promote the achievement of the strategic objectives. Therefore, and according to Ceitil (2007, p. 113), "The challenge becomes how to grab and prepare the people who will help achieve the results, identifying the skills that will ensure success in the medium and long term." In this sense, once again, the importance of investment in human capital, which Ceitil (2007, p. 113) considered as the "secret of success," is reinforced, as well as the needed competencies to achieve success. In this way, we are confronted with HR competency issues reason why it is important to know *what is the meaning of competence; how to acquire these competencies; and how to develop HR and their competencies.*

In what concerns competencies, many are the authors that have been studied it, namely Prahalad and Hamel (1990), Stoof et al. (2002), Sanchez (2004), Gilbert (2006), and Boyatzis (2008). It is also possible to see that this issue has been addressed in different contexts and field of analysis, such as sociology, management, education, among others.

Started research on competencies in the 1970s by David McClelland, and although the high number of definitions of the term that exists, competence can be understood as the set of capacities, attitudes, knowledge, behaviors, experiences of different natures and typologies that lead and help a collaborator to achieve his goals and improve his performance (Price 1997; Sanchez 2004). Also, Stoof et al. (2002), Sanchez (2004), and Ceitil (2007) argued that an individual's competence consists of skills, knowledge and abilities, experience, attitudes and behaviors, and many others that allow a collaborator to perform a certain function in an organization.

New organizations and competitiveness challenges require new skills, which a few years ago did not even exist. Currently, it is likely that employees will do (and do more) than is expected. It is necessary to bring together the skills and capacities to perform not only basic activities, but also to gather skills that previously were only assigned to leaders, such as problem-solving skills, initiative, critical thinking, and others. Employees have to invent, build, and innovate (Boterf 2005). Thus, today, being competent is increasing, being able to manage complex and unstable situations (Boterf 2005).

As a result of this statement, to be competent is to gather a set of skills that allow to manage unforeseen, complex situations, that people must own and that are difficult to imitate by the competition. These competencies (abilities and skills) were addressed by Prahalad and Hamel in 1990 through the publication of their article "The core competence of the corporation" in the Harvard Business Review, who called them core competencies, that is, core competencies that allow differentiation in organizations and are difficult to imitate.

Given that in any activity sector, industry, or service, HR must be competent at any organizational level, from top management to operators, and be equipped with different types of skills and are difficult to imitate, it can be questioned, *how to acquire competencies and core competencies?*

In order to ensure core competencies and the required knowledge to make organizations more competitive, it is critical that organizations develop HRM strategies, including HR training and development policies and practices. This way, it will be possible to acquire and develop competencies, using various tools and methodologies like training and development. As Santamaría et al. (2012) mentioned, training is a key aspect in the development of HR because when HR are trained, it allows them the exchange of experiences, the recycling of knowledge, the access to new topics and new tools, which lead to an increase in competencies capacity and, consequently, to a greater capacity of the organization.

2.4 Human Resources Qualification and Creativity

Having trained and competent HR requires the definition and implementation of HRM policies related to organization's strategy. This coordination of strategies can allow the company to overcome and face the constant challenges and anticipate new products, customers' needs, guaranteeing HR adequate to the market demand, with knowledge and skills.

As Barney (1991) argued, human capital is essential to the development of any business, being essential to manage them in order to enhance the application of their competencies. It is not enough to have technology, it is necessary to have the adequate intellectual capital with technical capacity, competencies, and knowledge that promote creativity (ability to create, develop, and be creative), and this is achieved with sustainable and planned policies of HRM.

In order to innovate in organizations, it is fundamental to have a culture of innovation that allows and conceives that failures (failure to achieve expected results) are part of the success of the innovative activity, since not always innovation projects are successful, even with trained, qualified, and creative collaborators. It is important to encourage experimentation to lead to new discoveries, either by creating or improving existing products and services, in order to respond in advance to market and customer needs with creativity. In this way, being creative can be part of the personality traits or be in the genes of a collaborator or can be the result of the application of knowledge obtained, either through professional experience or through the acquisition or sharing of knowledge and dynamics of the collaborator.

After all, what is the definition of creativity? What does creativity mean? What is the role of creativity in organizational innovation?

Creativity in organizations is currently a much discussed and addressed topic, although it is difficult to reach a consensus on how to measure it in organizations. It is an issue addressed both in vocational training institutions and academic institutions, as well as in the business and business environment, as a fundamental and essential element in the search for new solutions and new products, to meet the needs of customers and the market. In Portugal, there exist the Portuguese Association of Creativity and Innovation whose mission is to sensitize and promote projects in the area of creativity, although there are other private institutions that offer the market new approaches in the area of creativity, such as the Portuguese Association for Quality with workshops on the subject.

In Europe, in 2009, the European Commission established the European Year of Creativity and Innovation, with the importance of innovation-related creativity being reinforced as "solutions" to the exit of the deep crisis facing Europe.

According to Hennessey and Amabile (2010), the study of creativity in the last 20 years has grown and aroused the interest of researchers. To them, it has covered different areas, namely creativity in the context and perspective of the neurological and psychological aspects, the approach of creativity as an effect of the collaborators training, the influence of collaborators gender in the creative capacity, the influence of the organizational environment among other areas.

For Amabile (1996), creativity can be understood as the generation of new and useful products, practices, services, and procedures; it refers to the production of new and useful ideas. It follows that creativity is related to the creation of ideas that promote novelties.

Creativity, being the capacity to generate ideas and novelties, is, in individual terms, associated with people and their creative capacities, their characteristics of competence, personality, age, mental opening, and others, or it may be associated with the creative capacity of teams or work groups that develop together, exchange of ideas, experiences, and may lead to the triggering of new ideas (Binnewies et al. 2008; Hennessey and Amabile 2010; McAdam and McClelland 2002).

In organizations with high organizational culture, creativity has been a topic highly discussed requiring more and more people to be creative. In the entrepreneurial context, the concept of creativity is comprehensive, that is, creativity is related to new ideas, whether related to working methods or new products and processes. As people are an integral part of organizations, will creativity be an essential factor in creating new ideas and therefore will there be any relation between creativity and innovation in organizations? Innovation and creativity—are with the backs turned, are they complementary or is the former the consequence of the second?

2.5 *Innovation and Human Resources*

Seen as a current theme, much discussed in the business context, innovation is presented as a differentiating factor and with growing interest in the European economies, as presented in the 2020 Strategy. According to the European Commission (2014), this strategy is based on three priorities: smart growth, developing a knowledge-based economy and innovation, investing in HR education and research into new products and services; sustainable growth by promoting a balance between resource consumption and a more efficient economy; and inclusive growth, fostering employability by raising employment levels and reducing the poverty rate in Europe.

Nowadays, innovation and the organizations innovative capacity are factors that distinguish them leading to an international positioning that is based, in addition to other factors, on the intellectual capital that is in people (teams of motivated, trained, competent and creative employees with knowledge and skills) that allow the use of knowledge in order to add value.

Thus, innovation is crucial to competitiveness, and if companies fail to innovate, sooner or later they disappear. There are authors who have affirmed such importance, such as McAdam and McClelland (2002) and Solleiro and Castanón (2005), to whom competitiveness can be obtained, among other factors, through innovation activities and with employees with adequate training correspondent with their levels of responsibility; as well as Lau and Ngo (2004), who consider that innovation is an important indicator of competitiveness.

From the analysis of the literature, it was verified that the theme of innovation has been the subject of study and research by many authors, with many publications and papers in various scientific supports, from newspapers to magazines of the specialty, confirmed by the publication of Crossan and Apaydin (2010), which identified numerous authors and diversity of studies in the scope of innovation.

But what is the meaning of innovation?

In the current vocabulary, innovation is associated with something that adds value and that allows the differentiation of an organization from the competition. In order to drive innovation in an organization, it is required that it has a business culture with duly defined values, policies, and strategies and as an indispensable engine, committed top management with a long-term vision.

On the other hand, the concept of innovation is related to something new, a product, process or new method, that is, something that is new to the customer or the market or even to the organization, which can be called radical innovation. However, the definition also points to something that exists and can be significantly improved, leading us to what is called incremental innovation, that is, an existing product, process, or method that, with minor improvements, becomes an added value for the organization or for the market.

Thus, innovation is gaining more and more relevance in organizations, since an organization can be innovative not only in products but also in processes or modes of operation, as well as in new methods of promoting its services or products.

In this way, it can be inferred that any company can innovate, since even in organizational terms it can be possible, namely through the implementation of a new method of human resources performance appraisal, as well as a new way of training, or even implementing a human resource management system, existing currently a normative framework for this: NP 4427: 2003.

It is important to investigate whether the innovative capacity of an organization is conditioned by the competence and creativity of its HR and whether it facilitates innovation.

Associated with the innovation capacity of an organization, it is not only technology; it is also necessary to have high intellectual capital with capacities, competencies, and knowledge that promote creativity, the capacity to create, to develop, and to be creative, which is achieved with sustainable and planned HRM policies.

Solleiro and Castanón (2005) and Chen and Huang (2009) argued that HRM practices lead to HR as being an asset to be well-managed. This good management must allow that knowledge in the possession of HR can be used to create value in organizations, transforming value into a competitive advantage.

According to Chen and Huang (2009), studies developed by several authors considered that organizations must have HRM practices from a strategic point of view in order to influence employee behaviors and expectations and create value in the development of innovation, in order to obtain better results of the innovations. In this perspective, according to these authors, HR constitute a precious asset (value) in organizations.

Since HR are an integral part of an organization, they are embedded in an entrepreneurial culture, the reason why to promote innovation in organizations, it is essential to have a culture of innovation. This culture of innovation must allow and assume that failures (failure to achieve expected results) are part of the success of the innovative activity because innovation projects are not always successful, even with trained, qualified and creative employees. It is fundamental to respond in advance to the needs of the market and the customers, with creativity.

Being creative can be part of the personality traits and be in the genes of a collaborator or can be the result of the application of knowledge obtained, either through professional experience or through the acquisition or sharing of knowledge. Thus, existing HR knowledge and their application to value-added products and solutions must be well-managed. Knowledge management includes activities such as compilation, collection, analysis of knowledge, as well as dissemination and sharing in the organization. This knowledge can be cataloged and compiled into databases or shared folders on the intranet, often providing the self-development of other collaborators. This knowledge can be acquired through activities such as training or other forms of development and research.

In the European Union with the publication of the 2020 Strategy, it has undeniably been reinforced that innovation will gain ground and will not only be a fashion, but a necessity to combat competition from economies and countries with cheap and unqualified labor, or without environmental concerns or social responsibility.

In Portugal, in recent years, there have appeared documents and entities related to innovation such as the Manual of Best Practices in Innovation Management, edited by COTEC Portugal, as well as the publication of NP 4457: 2007, edited by the Portuguese Quality Institute. They allow organizations to implement innovation management systems, with the purpose of organizing and making their activities systematized, sharing knowledge. This standard was based on the Spanish standard UNE 16002: 2006, as well as other international documents recognized in the scope of innovation, such as the Frascati Manual and the Oslo Manual.

Once again, the role of HR in building an innovation system is reinforced which must have the competencies and responsibilities appropriate to the functions. Top management provides creative activities, through brainstorming, informal meetings, discussion forums, blogs, and others, as sources for continued and ongoing innovation. Can we then question, having competent HR an organization will be more innovative? What is the role of HR in the success of innovative companies? What is the position of Portuguese companies and what is the position of the Portuguese textile and clothing companies?

3 The Textile and Clothing Sector in Innovation and Human Resources in Portugal

In Portugal, in the last few years many ideas have arisen and many projects have been distinguished with merit, both at national and international level in various fields of research and innovation, proving that Portugal has talent. COTEC Portugal—Business Association for Innovation (for example), founded in 2003, has promoted and revolutionized innovation in Portugal, creating, among others, structures, working documents, tools to support innovation, application and orientation guides, self-assessment tools in the area of innovation, such as *Innovation Scoring*, and the "*SME Innovation COTEC-BPI*" prize, giving to the Portuguese companies the possibility to implement practices and strategies in the field of innovation and to recognize their innovation efforts.

On the other hand, this dynamic structure in the field of innovation, through the COTEC SME Innovation Network, encourages SMEs (small and medium-sized enterprises) to disseminate their add value, at the product or service level or by the capacity to generate value for society and for the country. This network of companies also encourages an exchange of experiences between the various companies from various sectors, with the aim of improving their performance.

The TCS is a traditional sector of the economy, which according to the ATP (Portuguese Textile and Clothing Association 2013) accounts for about 9% of Portugal's total exports, 20% of manufacturing employment, 8% of business volume in manufacturing, and 8% of manufacturing output. It is a sector made up of more than 7,000 companies in which 95% are SMEs and 85% are located in the northern region, with 51% of companies located in the Braga district (ATP 2013).

According to the CIP (n/d) study, the number of companies in the "textile manufacturing" sector is smaller than the number of companies in the clothing industry, being particularly SMEs (small and medium-sized enterprises Companies).

According to Mateus (2010), the TCS, is an integral part of a sector where there is a strong specialization, creating and realizing value-added products, either by the application of sophisticated techniques or by the application of innovative materials and raw materials. It is a sector that has focused on the innovation of products and processes, as the main competitive advantage, to survive in the market and surpass customer expectations, as well as to respond to the challenges of globalization, and thus move away from companies that are based in countries with cheaper labor, ensuring, this way, the continuity of its business.

As in any industry, human resources play a key role in the success of organizations. According to Fesete (2010), it can be seen that in the TCS human resources with higher education have been growing over the last years, while HR with basic education have been reduced, although the latter have still the greater representativeness.

It follows that in the context of HR and given the global situation, it is essential that they acquire training in different areas that promote differentiation and add value, such as innovation, marketing, and management control, being necessary to conduct new professional profiles with new competencies.

On the other hand, it can also be verified that there should be the reinforcement of competencies in the area of management and leadership for managers in order to promote and create an environment conducive to motivation and creativity (CIP n/d). According to the 2020 strategy, innovation and qualification of HR are priority areas to focus on in the near future.

4 Research Methodology

In this research, we have used a qualitative methodology, with semi-directive interviews, in companies belonging to the Portuguese TCS. These companies need to follow some conditions, namely to have a management system for R&D (research, development, and innovation), being certified and/or belonging to the COTEC SME Innovation Network. The main aim of the research was to verify what practices exist in the TCS, regarding the HRM and its implication in the success of its innovative capacity.

4.1 Research Questions

The development of this research was based on a macro-objective: to evaluate the relationship between the innovative capacity of organizations and their HRM practices, in order to answer the research basic question:

Will innovation in TCS companies achieved through qualified and trained HR? and what is their contribution to innovation?

From the literature review, it was verified that there are studies that investigated the relationship between HRM practices and value added, in several areas, including competitiveness, the organization's economic performance, innovation, age, and experience of HR, etc. The conclusion is common: HR play a key role in gaining value and contributing to the competitiveness of an organization, regardless of its size or sector of activity. Since innovation is one of the differentiating factors of the current organizations to compete in the global market, and in particular the organizations of the TCS, and by the literature reviewed, it is important to question:

1. *Do the companies considered most innovative in the TCS have techniques and practices of HRM, with competent HR and continuous training practices?*

Based on studies collected from Fesete (2010) and CIP (n/d) about the characterization of HRM practices, textile and clothing companies are essentially SMEs and do not promote the practice of true HRM. Based on the selected sample, this question will be evaluated, in order to confirm whether this is applied and practiced in innovative companies.

Another relevant research question is the competencies of employees and their relationship with the capacity to innovate, asking the following question:

2. *Are the competent and qualified employees who promote and induce business innovation?*

By reviewing the literature, it was verified that competencies are essential in any situation and in the innovation aspect, and there are characteristics/new competencies that HR must possess to leverage innovation in organizations, as advocated by Siengthai and Bechter (2001), Boterf (2005), and Chen and Huang (2009). The management practices of training and qualification are essential to the acquisition and development of HR competencies. It is therefore necessary to verify whether TCS companies have constant and regular training practices that promote the acquisition of new competencies.

Another relevant issue and target of the study is the HR creativity, in the context of innovation in the TCS. Thus, one may question:

3. *Do textile and clothing companies promote and stimulate their HR creativity in order to obtain new ideas and suggestions? Who are the most creative in the organization? Creativity has been the subject of much research in several perspectives; however, within the scope of this research, it is intended to evaluate which practices are used, which tools and if they are applied.*

Last but not least, it is important to validate the results of innovation in the TCS companies and to what extent HR have an impact on the success of this innovation. Thus:

4. *Is innovation the result of chance or the result of systematized practices and stimulation of creativity of qualified HR?*

4.2 Sampling and Data Collection Tool

We chose to focus the study on companies in the Portuguese TCS, choosing a sample for convenience. In what concern the procedures to be followed in methodological terms, we chose the qualitative method, using interviews to obtain data on the research topic. Taking into account the theme, it is important to listen to the various interlocutors, their practices and experiences about HRM, their creativity, and the innovation in their organizations.

Considering the research purpose and questions, semi-structured interviews were conducted, based on a previously drawn script, which looked to cover the issues related to research and the theoretical framework/literature review.

We interviewed eight employees from six companies (A, B, C, D, E, F) that accepted to participate in the research, being essentially responsible for the area of innovation (four—from A, B, C, D companies), Top Managers (two—from D and F companies), financial area director (one—from E company), and the responsible for the area of human resources (one—from A company). The interviews took place at the premises of the company, being conducted between September 9, 2013, and March 13, 2014, depending on the availability of the companies, having an average of 1 h and 30 min. It should be noted that the interviews were recorded. The objective of these interviews was to collect "in loco" a wide range of information, including opinions, practices, in order to substantiate the findings of the investigation.

5 Results Analysis and Discussion

In order to facilitate the presentation and understanding of the results obtained, it was considered the division of the research themes into different categories, which in turn were divided into several dimensions of analysis, as observed in Table 1.

For each dimension of analysis, the questions related to them were worked, as well as the responses given by the respondents.

For reasons of text limitation, in this paper we will only present a synthesis and conclusions of the analysis of the various categories and dimensions of analysis.

As we have seen before, based on a selected sample, this research aims to study the practices and techniques used in innovative organizations in order to answer the research basic question:

Is innovation in TCS companies achieved through qualified and trained HR? And what is their contribution to innovation?

Table 1 Classification of categories and dimensions of analysis

Category	Dimension of analysis
HRM	Concept and HRM practices in TCS
	Competencies and policies of competencies assessment
	Training and training policies
Creativity	Practices to stimulate creativity and tools used
	Conditions for creativity in organizations
	competencies and personality traits of employees
Innovation, HR, and results	Capital gains from innovation
	The role of HR in innovation
	Internal conditions of innovation

In order to elucidate and conclude on this issue, other research questions have been established, to which, and after the interviews analysis, were obtained the following results:

1. *Do the companies considered most innovative in the TCS have techniques and practices of HRM, with competent HR and continuous training practices?*

By analyzing the respondents answers to the various questions formulated about HRM practices and employee competencies, it can be noted that practically all the organizations in the sample have HRM practices, even if in day-to-day management (recruitment, training) and in legal issues (salaries, absenteeism, etc.). Thus, in order to TCS develop and innovate, it will be necessary for organizations to invest more in HR policies since HR are one of the pillars of any organization.

As one of the interviewees (company D) said, "*it is hard to say which is the magic wand to have the best HR, because it does not exist, but it is the result of a lot of management involvement.*" This statement suggests the importance of management in the definition of HR strategies, that are not always a priority, and management should include a budget to promote and improve the qualification of HR.

It was also observed that the companies of the selected sample have training practices, allowing the acquisition of new knowledge, skills, and competencies, in different hierarchical levels, from the administration to the employees, as well as in different thematic ones, being an added value in innovation, which leads to the conclusion that, in order to have good results and capable HR, training is essential, confirming the established in the Strategic Plan presented by ATP on September 25, 2014.

On the other hand, in these companies, in addition to recruitment, training and legal compliance activities, it is not common to have other HRM activities, such as competencies and performance appraisal and career plans, the reason why it may be suggested to improve these practices.

2. *Are the competent and qualified employees who promote and induce business innovation?*

All respondents indicated competence and certain personality traits of employees as a determining factor for innovation success. They also indicated that the key competencies of employees are positively related to innovation, not only technical knowledge, but also social skills and know-how to be, attitudes and behaviors, which are fundamental in an organizational culture of innovation. It is important to highlight attitudes and competencies in the areas of teamwork, learning ability, open spirit, openness to change, proactivity, creativity, among others.

However, the evaluation of these competencies is carried out informally and as they are SMEs, the proximity between management/administration and employees is high, facilitating the evaluation process. The way in which key competencies are assessed can be rethought.

Albeit, there is a positive relationship between competence, qualification, and innovation, in which collaborators with knowledge, with different experiences and even academic degree, induce innovation in organizations.

Although the studied companies have some HRM practices, TCS companies will still have a way to go in terms of HR qualification, competencies, and training. This is mentioned by the studies of Fesete (2010), CIP (n/d), and the Strategic Plan for TCS (Dinis et al. 2014), thus reinforcing the need for TCS organizations to increase and reinforce HRM practices to meet current and future challenges.

3. *Do textile and clothing companies promote and stimulate their HR creativity in order to obtain new ideas and suggestions? Who are the most creative in the organization?*

Amabile et al. (1996) reinforced the importance of organizations practicing creativity as a starting point for innovation, making creativity a prerequisite for innovation to happen. And because creativity is a person characteristic, there are those persons who, in groups or in isolation, give ideas and create or recreate new or significantly improved products, methods, and practices.

The theme of creativity, as explained in the literature review, is very complex and comprehensive, and from the interviewees' answers, it is concluded that any collaborator in an organization can be creative, regardless of the qualification he possesses and his position in the hierarchy, although he has to gather competencies and personality traits that lead to creativity. That is, not always a TCS collaborator, with a higher academic degree, is more creative and contributes more to innovation. However, qualification, in terms of a higher academic degree, favors creativity and, therefore, innovation.

The obtained results allowed us to observe that the studied organizations confirm the importance that HR must have a set of skills and abilities that enable them to be competent, being necessary to reinforce their autonomy, openness, ability to learn and the will to learn, creative and imaginative capacity, problem-solving capacity, commitment, dedication, organization, responsibility, autonomy, honesty, and creativity. Another relevant key competency in the context of innovation is the HR ability in the interaction and integration of work teams so essential to innovation management. As stated by the interviewee of company C "*it is very important to know how to be with the rest of the group, a person can be very good at working alone, but if you can not get the message to the rest of the group and work as a group, you will never be competent. Also, being proactive and always anticipating is also very important.*" This statement confirms the importance of HR in the innovation area demonstrates and possesses competencies in knowing how to be, as referred by Ceitil (2007).

It is therefore necessary to assess whether innovative firms use competencies and creativity by encouraging them to stimulate HR in the production of new ideas and solutions. From the respondents' answers, it can be concluded that there are practices to foster creativity through meetings, with or without defined frequency, as witnessed by the interviewees of the companies, B, C, D, and E. In these meetings are presented and discussed ideas and problems, which, by their nature, may give rise to new products, services, methods, and others. Frequently using brainstorming, work meetings with multidisciplinary teams are productive and work as a privileged tool for sharing ideas and knowledge. This is confirmed by the interviewee of company D when he says that "*there is an innovation team that involves a group of people from different areas, administration, research, development and innovation department, commercial, more technical collaborators; and in this team we try to gather key competencies, so that this creativity responds to all department needs, so that the knowledge is more open to all.*"

There have been listed the competencies and personality traits essential to innovation and that are valued by innovative organizations. Based on them, it is advisable to either collaborators who work or will work in TCS and non-innovative organizations, to invest in this issue, since in TCS more and more these concerns must be increased with a view to innovation.

However, there has been a tendency in which senior managers are referred to as the most creative, the ones who give the most valuable ideas. Such practice can occur, as senior managers gather more knowledge, more training and can gather personality traits, which stand out in relation to operational employees. It should be noted, however, that intermediate managers and operational collaborators, without academic qualification, but with professional experience and with certain personality traits can make innovation happen in an organization, as referred to and reinforced by companies A and B. This reveals that, by the studied sample, not always qualification as an academic degree leads to valuable ideas that translate into add value in innovation, but other factors that condition creativity, as advocated by Binnewies et al. (2008), where the schooling level may not increase creative capacity. This depends on other factors, such as age, the environment in which the employee is inserted, the role that he/she performs, the professional experience, the knowledge acquired, that is, there is no direct relationship between creativity and HR qualification, in the TCS.

4. *Is innovation the result of chance or the result of systematized practices and stimulation of creativity of qualified HR?*

Alves et al. (2007) concluded that innovation occurs when there are HR with appropriate qualifications. However, Amabile et al. (1996) pointed out that successful innovation also depends on other factors, that is, not only from creative ideas originated within the organization, but also from original ideas, from any side (such as technology transfer). This realization is based on the fact that innovation depends on other factors that not only HR, nor their creative ideas.

Taking into account these positions, and in what concerns the studied organizations the obtained answers allowed us to conclude that for innovation happen it is necessary to create and have several conditions, including qualified HR. However, the success of innovation in the studied organizations does not identify a particular human resource or a specific qualification, referring to success as the result of teamwork with multiple skills and competencies. At the same time, all the interviewees point out a very strong relationship between qualified HR and innovation, although they also point out that creativity can be of any collaborator, any hierarchical level and with any academic degree.

HR qualification and training are identified as priority investments in TCS organizations in order to facilitate innovative capacity, although there are also barriers to innovation such as HR resistance to change and the lack of creativity.

Thus, we can conclude that the employees' competencies, the practices and stimuli of creativity, and the stimulus to learning and experiment encourage innovation. At the same time, it can be observed that TCS companies present different typologies of innovation, such as the definitions of the Oslo Handbook introducing new products, value-added products, new processes and technologies, new working and communication methods, marketing, obtaining added value to their organizations, providing the necessary conditions to assert themselves in global markets and competing with differentiation in emerging markets.

6 Final Considerations

Based on the obtained results, it should be noted that although the studied organizations do not apply HRM practices, in their global scope, they have good practices that allow them to achieve innovation success, with proximity and a commitment of different hierarchical levels and with different qualifications and academic degrees. Also based on the obtained results, we must mention the importance of creativity as a fundamental element for innovation in TCS organizations, whether with internal collaborators or with partnerships. In addition, it is important to reinforce the acquisition of knowledge about new creative techniques and apply them with the aim of improving the creative capacity.

From the results obtained and in order to answer the question originally formulated:

Will be the innovation in TCS companies achieved through qualified and trained HR? And what is their contribution to innovation? It can be concluded that the studied organizations, with qualified HR and other conditions, such as financial, technological, a network of partners, among others, manage to induce and make innovation happen. This is not the result of chance, but rather the result of systematized practices of HRM, with capable HR and with certain key competencies, with collaboration and motivation, as well as discussion and stimulus of ideas generation, with a view to innovation.

Thus, TCS organizations must continue to "make the leap;" bet on market niches; bet on innovation, as a differentiating factor from other emerging economies' organizations, as suggested by Vaz (2014); and investing in HR creativity and qualification techniques, in order to project TCS companies to levels of superior performance and for a more promising future for TCS.

References

Alves, J., Marques, M. J., Saur, I., & Marques, P. (2007). Creativity and innovation through multidisciplinar and multisectoral cooperation. *Creativity and Innovation Management, 16*(1), 27–34.

Amabile, T. M. (1996). *Creativity and innovation in organizations*. Harvard Business School. Available at http://cms.schwarzpharma.com/_uploads/media/7165_Amabile%20Creativity%20and%20Innovation%20in%20Organizations.pt. Accessed September 12, 2013.

Amabile, T. M., Conti, R., Coon, H., Lazenby, J., & Herron, M. (1996). Assessing the work environmental for creativity. *The Academy of Management Journal, 39*(5), 1154–1184.

ATP—Associação Têxtil e Vestuário de Portugal (2013). *Guia de orientação para a Inovação e Empreendedorismo do Cluster Têxtil Moda nos Vales do Ave e Cávado*. Portugal: ATP.

Auw, E. (2009). Human capital, capabilities & competitive advantage. *International Review of Business Research Papers, 5,* 26–36.

Barney, J. (1991). Firm resources and sustained competitive advantage. *Journal of Management, 17*(1), 99–120.

Becker, B., & Gerhart, B. (1996). The impact of human resource management on organizational performance progress and prospects. *The Academy of Management Journal, 39*(4), 779–781.

Binnewies, C., Ohly, S., & Cornelia, N. (2008). Age and creativity at work: The interplay between job resources, age and idea creativity. *Journal of Managerial Psychology, 23*(4), 438–457.

Boterf, G. L. (2005). *Construire les compétences individuelles et collectives: Les réponses à 80 questiones*. Paris: Éditions D'Organisation.

Boyatzis, E. R. (2008). Competencies in the 21st century. *Journal of Management Development, 27*(1), 5–12.

Buckley, J. P., Christopher, L. P., & Prescott, K. (1988). Measures of international competitiveness: A critical survey. *Journal of Marketing Management, 4*(2), 175–200.

Buller, F. P., & McEvoy, M. G. (2012). Strategy, human resource and performance: Sharpening line of sight. *Human Resource Management Review, 22,* 43–56.

Ceitil, M. (2007). *Gestão e desenvolvimento de competências,* 1st edn., 2nd printing. Lisboa: Edições Sílabo.

Chen, C. J., & Huang, J. W. (2009). Strategic human resource practices and innovation performance: The mediating role of knowledge management capacity. *Journal of Business Research, 62,* 104–114.

CIP—Confederação Empresarial de Portugal. (n/d). *Estudo de caracterização dos recursos humanos no setor têxtil e do vestuário*. Available at http://cip.org.pt/wp-content/uploads/2014/10/Estudos-sobre-os-RH-da-ITV.pdf. Accessed September 30, 2014.

Commission, European. (2014). *Innovation scoreboard 2014*. Belgium: European Commission.

Crossan, M. M., & Apaydin, M. (2010). A multi-dimensional framework of organizational innovation: A systematic review of the literature. *Journal of Management Studies, 47,* 1154–1191.

Dessler, G. (2013). *Human Resource Management* (33rd ed.). Harlow: Pearson Education Limited.

Dinis, A. P., Agis, D., & Vaz, P. (2014). *Plano Estratégico Têxtil 2020: Projetar o desenvolvimento da fileira têxtil e do vestuário até 2020*. Portugal: ATP.

European Commission. (2011). *Competitiveness Report.* Available at http://ec.europa.eu/research/innovation-union/pdf/competitiveness-report/2011/executive-summaries/pt.pdf. Accessed January 23, 2013.

European Commission. (2013). *Competitiveness Report.* Available at http://ec.europa.eu/research/innovation-union/pdf/competitiveness_report_2013.pdf. Accessed January 23, 2013.

Farole, T., Reis, J. G., & Wagle, S. (2010). *Analyzing trade competitiveness: A diagnostics approach.* The World Bank. Available at http://econ.worldbank.org/external/default/main?pagePK=64165259&piPK=64165421&theSitePK=469382&menuPK=64216926&entityID=000158349_20100601154408. Accessed January 10, 2013.

Fesete. (2010). *Ponderação das categorias profissionais nas grelhas das indústrias Textil, Vestuário e Calçado.* Porto: Fesete.

Gilbert, P. (2006). Competence: Travelling concept, fixed meanings. *Psychologie du Travail et des Organizations, 12,* 67–77.

Henessey, B. A., & Amabile, T. M. (2010). Creativity. *The Annual Review of Psychology, 61,* 569–598.

Huselid, A. M. (1995). The impact of human resource management practices on turnover, productivity, and corporate financial performance. *Academy of Management Journal, 38*(3), 635–672.

Lau, C. M., & Ngo, H. Y. (2004). The HR system, organizational culture, and product innovation. *International Business Review, 13,* 685–703.

Maier, A., Brad, S., Nicoara, D., Maier, D. (2014). Innovation by developing human resources, ensuring the competitivenesss and success of the organization. *Procedia—Social and Behavioral Sciences, 109,* 645–648.

Mateus, A. (2010). 'Indústria Portuguesa—Situação atual e evolução recente'. *Apresentação CIP.* Available at http://www.google.pt/url?sa=t&rtc=j&q=&esrc=s&source=web&cd=1&ved=0CB8QFjAA&url=http%3A%2F%2Fwww.acif-ccim.pt%2FAdmin%2FPublic%2FDownload.aspx%3Ffile%3DFiles%252FFiler%252FACIF%252FPaginas%252Festudo-industria-portuguesa.pdf&ei=HZICVIPbFMrcasjagbgM&usg=AFQjCNHbxfrKoNhjywAS6-NDpPIPOZui-Q-. Accessed July 24, 2014.

McAdam, R., & McClelland, J. (2002). Individual and team based idea generation within innovation management: organizational and research agendas. *European Journal of Innovation Management, 6*(2), 86–97.

Noe, A. R. (2010). *Employee training and development.* New York: McGraw Hill.

OECD. (2010). Innovation to strengthen growth and address global and social challenges. *Ministerial Report on the OECD innovation strategy.* Available at http://www.oecd.org/sti/45326349.pdf. Accessed January 23, 2013.

Prahalad, C. K., & Hamel, G. (1990). The core competence of the corporation. *Harvard Business Review, 68*(3), 79–91.

Price, A. (1997). *Human resource management in a business context.* International USA: Thomson Business Press.

Sanchez, R. (2004). Understanding competence-based management identifying and managing five moves of competence. *Journal of Business Research, 57,* 518–532.

Santamaría, L., Nieto, J. M., & Miles, I. (2012). Servisse information in manufacturing firms: evidence from Spain. *Technovation, 32*(2), 144–155.

Siengthai, S., & Bechter, C. (2001). Strategic human resource management and firm innovation. *Research and Practice in Human Resource Management, 9*(1), 35–57.

Solleiro, J. L., & Castanón, R. (2005). Competitiveness and innovation systems: the challenges for Mexico's insertion in global context. *Technovation, 25,* 1059–1070.

Stoof, A., Martens, L. R., Merrionboer, V. G. J. J., & Bastiens, T. J. (2002). The boundary approach of competence: A constructivist aid for understanding and using the concept of competence. *Human Resource Development Review, 1*(3), 345–365.

Ulengin, F., Kabak, O., Onsel, S., Aktas, E., & Parker, B. R. (2011). The competitiveness of nations and implications for human development. *Socio-Economic Planning Sciences, 45,* 16–27.

Vaz, P. (2014). Temos plano e estratégia para 2020. *Jornal Têxtil CENIT, ano XVII, 187,* 14–15.

World Economic Forum. (2014). *The global competitiveness report 2014–2015* (Full ed.). Geneva: Word Economic Forum.

Index

C. Machado and J. P. Davim (eds.), *Entrepreneurship and Organizational Innovation*, Management and Industrial Engineering, https://doi.org/10.1007/978-3-030-19289-1

Printed by Printforce, the Netherlands